The Polluters

BENJAMIN ROSS

STEVEN AMTER

The Polluters

The Making of Our Chemically

Altered Environment

OXFORD
UNIVERSITY PRESS

2010

OXFORD
UNIVERSITY PRESS

Oxford University Press, Inc., publishes works that further
Oxford University's objective of excellence
in research, scholarship, and education.

Oxford   New York
Auckland   Cape Town   Dar es Salaam   Hong Kong   Karachi
Kuala Lumpur   Madrid   Melbourne   Mexico City   Nairobi
New Delhi   Shanghai   Taipei   Toronto

With offices in
Argentina   Austria   Brazil   Chile   Czech Republic   France   Greece
Guatemala   Hungary   Italy   Japan   Poland   Portugal   Singapore
South Korea   Switzerland   Thailand   Turkey   Ukraine   Vietnam

Published by Oxford University Press, Inc.
198 Madison Avenue, New York, NY 10016

www.oup.com

Oxford is a registered trademark of Oxford University Press.

Library of Congress Cataloging-in-Publication Data
Ross, Benjamin.
The polluters: the making of our chemically altered
environment / Benjamin Ross, Steven Amter.
    p.   cm.
Includes bibliographical references and index.
ISBN 978-0-19-973995-0
 1. Chemical industry—Environmental aspects—United States.
 2. Chemicals—Environmental aspects—United States.
 3. Pollutants.
 I. Amter, Steven.   II. Title.
TD195.C45R67 2010
363.7380973—dc22   2009045748

9 8 7 6 5 4 3 2 1

Printed in the United States of America
on acid-free paper

ACKNOWLEDGMENTS

THIS BOOK IS THE fruit of ten years of research into the history of pollution, and it has come into existence only because so many people generously shared their time and knowledge along the way.

When we first began our research into the history of pollution, Peter Skinner and Craig Colten very generously allowed us to search and freely copy the document collection they assembled when they wrote their pioneering book *The Road to Love Canal*. Their advice and encouragement has continued without fail.

Many other people have assisted our research over the years. Three who went out of their way to help us deserve special mention. Leon Billings allowed us to use his extraordinary collection of printed congressional hearing records. Travis Wagner shared his deep knowledge of hazardous waste regulation and its history. Anthony Travis unhesitatingly and unfailingly offered us his encouragement, shared his knowledge, and, most important, gave wise advice.

Other historians generously took time to discuss issues with us, including Alan Derickson, Alan P. Loeb, David Michaels, John Parascandola, Frederick Quivik, and Lynn Snyder. Charlotte Sellers of the Jackson County, Indiana, public library went far out of her way to help our research into the Sayers family. We also gained valuable insight and information from relatives and friends of people who figure in the book. We thank Molly Gosline and Anne Barnes, nieces of Carl Gosline; Jacqueline Tarr Dempsey and Douglas Janney, daughter-in-law and grandson of Omar Tarr; C. Kenneth Banks, Jr.; and Joyce Zimmerman and Margaret Budd, friends of George Best. Thomas Burke helped us with his own recollections about looking for solvents in ground water in the 1970s.

The idea of writing this book grew out of research we undertook for attorneys engaged in litigation over the responsibility for environmental damage and the costs of cleanup. As we repeatedly encountered the same people, the same organizations, and the same behavior in different contexts, we concluded that there was a story that had never been brought together. This study would never have been written were it not for the confidence these attorneys placed in us. Particular thanks go to Jack Atkins, Stuart Calwell, Elizabeth Crook, Kevin Hannon, Anthony Jenkins, Kevin Madonna, Steve Medina, Ken Newa, Mike Papantonio, Gary Praglin, Tod Robins, Victor Sher, Ron Simon, John Skaggs, Donald Stewart, Mary Jane Thies, and Duane Westrup.

Numerous archivists assisted our research, in person or at a distance. They include Dennis Bilger and Randy Sowell of the Truman Presidential Library, Judy Grosberg of the National Cancer Institute Director's Office, Nancy Miller of the University of Pennsylvania Archives, Patrick Shea of the Chemical Heritage Foundation, Brian Shovers of the Montana Historical Society, and Paul Theerman of the National Library of Medicine. Archivists who helped us locate and identify photographs include Barbara Harkins of the NIH History Office, Cindy Lachin of the FDA History Office, Delores Morrow of the Montana Historical Society, Christie Peterson of the Muskie Archives at Bates College, Daniel Whittemore of Sheppard Powell Associates, and Leila Wiles of the Izaak Walton League. We also thank Kate Barry for her help as a research assistant.

Finally, the writing of this book would have been utterly impossible were it not for the many historical researchers in this field who came before us. A history of this scope cannot possibly be written entirely from primary sources. Only after years of painstaking investigations of individual episodes is there any hope of bringing together a synthesis. We have relied on numerous published monographs, and we are especially indebted to three authors of unpublished dissertations, Marvin Brienes, Lynn Snyder, and Adelyne Whitaker. Even where we disagree with their interpretations, we owe them an immense debt of gratitude.

CONTENTS

The Polluters

The first victim of the Donora smog, Peter Starcovich, is laid to rest as factory smoke rises in the distance. (Photograph by Alfred Eisenstaedt. Courtesy of Getty/Time and Life Pictures.)

| The Sorcerer's Apprentices

Uncontrolled one, born in Hell

Will you drown our house entire

In the flood already streaming

Out every door and windowsill?

  Corrupted broom that will not heed,

  Be lifeless stick again, I plead!

—J. W. von Goethe, "The Sorcerer's Apprentice"[1]

DONORA, PENNSYLVANIA, BURIED ITS dead on Election Day. On Tuesday morning, November 2, 1948, the sun shone brightly on a gritty mill town that just two days earlier had emerged from a siege of poisonous smoke. For four days, stagnant air had trapped smelter fumes in the steep-sided valley where houses nestled alongside factory buildings. As the air grew thick with smoky fog, it took an effort just to breathe, and many sought refuge on higher ground. On Sunday, when the air at last cleared, nearly half the population was ill and 20 victims lay dead or dying. The homes of the dead were clustered around the zinc works, one of two metal plants that sustained the community.

Since Saturday, the town had been sending urgent calls for help to Washington. The appeals were rebuffed. On Tuesday morning, an official of the

Public Health Service repeated the refusal, telling newspapers that the disaster had been nothing more than an "atmospheric freak."

Wednesday morning brought astonishing news about the election. Republican presidential candidate Thomas Dewey, the overwhelming favorite, was defeated by President Truman, and the Democrats took control of Congress. By that afternoon, the Public Health Service made a sudden U-turn and agreed to investigate the disaster. On Thursday morning federal scientists started to arrive in Donora.

Seven months later, on June 7, 1949, Governor Earl Warren of California stood before a news conference in Sacramento. He was there to denounce water pollution legislation that the state Assembly had just passed and sent to the Senate. The bill, he charged, would put the polluters in charge of the pollution control setup. The director of the state health department stood beside him, warning of a plague of bad water that he likened to the smog in Los Angeles. In the days that followed, a compromise was negotiated. When the law was enacted two weeks later, the most objectionable clauses were gone.

As these events faded from the headlines, the friends of clean air and clean water were to all appearances the victors. But once the spotlight of public attention turned elsewhere, their hard-won gains proved phantom. The polluters had at their disposal a battery of weapons—political, economic, and scientific—forged by the chemical industry and its allies. In struggle after struggle over the preceding decades, business interests had preserved for themselves the freedom to foul their surroundings. The weapons that had won these battles were deployed again in Donora and Sacramento, and they once more proved their worth. The Donora investigation was sidetracked; the Public Health Service failed to answer—or even ask—the central questions about the causes of the catastrophe. The new California law enshrined manufacturers' right to discharge waste, creating toothless Water Pollution Control Boards with powers so circumscribed that they had little choice but to ratify what industry decided to do.

What was the origin of the armament unsheathed in Donora and Sacramento? Under what conditions and for what purposes was it forged? Was it employed as its creators intended, or did it escape their control like the broom conjured by the sorcerer's apprentice? These are vital questions for understanding today's environmental dilemmas—and beyond that, they evoke some of the most fundamental problems in social thought. How does economic power influence government? What is the basis of scientific authority? Is science value-free, or is it shaped by social and economic conditions? For more than a century, thinkers and scholars have debated such questions.

Political writers of all schools recognize that private economic interests can turn government policies to their advantage. Opinions differ mightily, however, in their understanding of how that happens and in their prescriptions for improvement. What a libertarian describes as rent seeking is called regulatory capture by a pluralist; a Marxist explains it by way of economic determinism.

Debates over such basic questions of political theory will surely long continue, but one can at least hope for more empirical investigation and less falling back on abstract reasoning. This book will perhaps contribute toward that end. Early struggles over environmental control offer a striking case study in the relationship between business and government. Politics, pollution, and science came together in a way that foreshadowed the technological complexity of today's governance. The story opens a window into the broader issues of political power and scientific knowledge even as it illuminates current environmental conflicts.

———

Wastes were a problem from the earliest days of chemical manufacturing. But the American chemical industry of the nineteenth century lagged far behind its European competitors, and the emissions from its factories drew little notice beyond their immediate surroundings. The First World War brought a sudden change, with larger plants and new synthetic chemicals that bore little resemblance to the original raw materials. Effluents quickly grew in volume and diversified in content. As pollution worsened and new problems emerged in the course of the 1920s and 1930s, scientists and the public increasingly saw the need for control and demanded action.

Leaders of the industry recognized the need for cleanup, but they were allergic to government oversight. Chemical companies insisted on doing things themselves, at their own pace, with their own means, and they gathered their forces for the fight to keep the government out. An armament of methods was developed to fend off outside pressure. One of the industry's common tactics can be summed up as "spill, study, and stall." When outside pressure to do something about pollution became strong, a study of the problem would be launched as an alternative to expensive action. The study would be carried out by the polluters themselves or, if it was feared that a blatantly self-serving study would lack credibility, under their influence.

Research was directed most often toward devising new techniques for cleaning up wastes. The chemical industry placed enormous faith in technological progress. It certainly desired innovative solutions to pollution problems, especially solutions that were profitable or at least inexpensive. On occasion, such discoveries were made. Yet the more immediate—and often sole—consequence of research was to buy time by deflecting the demand for control when technology was already available.

Another type of study tried to clarify whether, or at what levels, a pollutant was harmful. Research of this kind was more dangerous for manufacturers because it posed a great risk. Chemicals might be shown to be highly toxic or even to cause cancer; such discoveries could trigger an irresistible demand for control. The first line of defense against this threat was to stop investigations before they began. Chemicals, the industry maintained, were innocent until proven guilty, so lack of knowledge justified lack of control.

There was a second line of defense against the threat that greater knowledge would bring bad news. When study could not be avoided, friendly researchers would offer a predetermined conclusion. They would cherry-pick data, design experiments to give a desired answer, or sometimes offer reassurances backed by nothing more than the sheer force of assertion. The exercise of political, financial, and public relations muscle would turn this into "authoritative science," often in the face of criticism from scientists of much greater attainment. During the Republican administrations of the 1920s, the federal government could be called on to provide such studies. With the coming of the New Deal, government was no longer a faithful servant of business. New private-sector institutions, ostensibly neutral scientific bodies but controlled behind the scenes by industry, were created to supply obedient expertise. A series of decisions that would have enormous consequences for public health—about leaded gasoline, black lung, DDT, air and water pollution—were justified by this technique.

As the Second World War approached, and even more once it had passed, the dangers of unchecked industrial pollution were widely sensed. Engineers and managers inside the chemical industry knew far more than the public, and the leaders of major companies like DuPont and Dow understood the need for action. But the extreme laissez-faire political beliefs of the industry's top management led them to reject federal environmental controls. They insisted that environmental control should be a state and local matter, leaving regulators hobbled by political pressure and the threat that factories could relocate to friendlier jurisdictions. Absent effective outside regulation, the industry would make its own decisions.

In large decentralized organizations, process can be policy. Within chemical companies, responsibility for environmental control fell to specialists who held staff positions. They often lacked the clout to overcome resistance from operating managers, whose incentives were driven by internal profit targets and outside competition. In the profit-driven marketplace, competition pushed all down to the level of the least scrupulous, who had at their disposal the apparatus created to fight off outside interference.

Matters came to a head in the years just after the Second World War. Contamination problems were magnified by an increased scale of production and made less tractable by an avalanche of new synthetic chemicals. Influential scientists and public officials saw a need for federal regulation of air, water, and pesticides. A political and bureaucratic struggle ensued—sometimes in the open, more often behind the scenes—through which the chemical industry preserved for itself the right to determine what would be emitted from its plants. Industry's victory was codified in federal laws—the Federal Insecticide, Fungicide and Rodenticide Act of 1947, the Water Pollution Control Act of 1948, and the Air Pollution Control Act of 1955—that belied their reassuring titles by rejecting federal regulation in three major domains of environmental policy.

Techniques had been developed, habits formed, laws passed, institutions created. By now the laws, habits, and institutions had taken on a life of their own. Like the water carrier summoned by the sorcerer's apprentice in Goethe's poem, they continued their work even after the intentions of their creators had been fulfilled. The leading corporations that had put the machinery into motion, the DuPonts, Union Carbides, and Standard Oils, watched as the spirit they had raised up sometimes exceeded its intended tasks. The conjurers had the power to break the spell, but the formula of federal regulation was anathema to them. For 20 years more, the chemical industry battled successfully against outside oversight. Engineers' valiant efforts at control, hobbled by the workings of an unregulated market, were outraced by a tide of pollution.

Despite postwar defeats, the public movement for environmental control never fully disappeared. Concern about the purity of food, stoked through the 1950s by congressional hearings and condemnations of unsafe food, led to incremental tightening of pesticide regulations. In 1962 came Rachel Carson's *Silent Spring*, first serialized in the *New Yorker* and then a best-selling book. Burning rivers soon vied with smoggy air for the media spotlight, and a liberal Congress turned its attention to pressing environmental problems. Earth Day in 1970 launched a national grass-roots movement. The next decade saw a flood of legislation, creating the national system of environmental regulation that industry had successfully fended off after the Second World War.

But the techniques employed years ago in Donora and Sacramento have never gone out of favor. Discovery of new environmental problems is discouraged, with research that might find them starved of funds. When alarming findings do emerge, well-paid advocates concoct grounds for doubt. Study follows on study as a substitute for action. Sixty years later, these strategies are still in use, protecting polluters who spew out toxic chemicals and globe-warming gases.

———

Chemical manufacturers loved to talk about the magic of chemistry. Fifty and one hundred years ago, their enterprise was at the farthest frontier of science and industry, breaking new ground as it transmuted wastes into valuable commodities, reinvented familiar materials with new dyes and coatings, conjured up synthetic fibers and plastics. Edward Collins, the *New York Times'* financial editor, summarized the industry's achievements when the American Chemical Society celebrated its 75th anniversary in 1951: "The story of modern chemistry is all the tales of the Arabian Nights retold, each in a multitude of variations, by a modern Scheherazade of infinite imagination." The cavalcade of chemistry's advance seemed so miraculous that the image of wizardry became a cliché of popular culture. "Modern chemistry rubs its Aladdin's lamp, shakes up its test tubes, and, presto!" began a *Washington Post* account of the industry's new wonders.[2]

Occult powers, in literature and in life, are not always benign. The Sorcerer's Apprentice was one magical tale that chemical companies chose not to recall. It is a story that had much to teach them.

PART I | Summoning the Spirits

CHAPTER 2 | Pollution Goes to Washington

...a few inconsequential matters in connection with the safety of human life

—Herbert Hoover, 1922[1]

L ONG, LONG BEFORE THERE was a chemical industry, humans feared for the purity of the air and water that sustain life. Alongside natural and supernatural terrors, the danger of poisoning by human agency was recognized early. The ancient authors Xenophon, Lucretius, Vitruvius, and Pliny describe polluted air and water around mines.[2] Under Edward I of England in 1306, a prohibition on the burning of coal in London was issued—and, it seems, promptly ignored.[3] Court cases on groundwater contamination are attested in Europe as early as 1349. In the wave of bigotry that followed the Black Death, Jews living in what is now Switzerland were accused of causing the disease by placing poison on the ground near wells. The judges ruled that the defendants must be guilty because they confessed after only very light torture.[4] Even in an era of medieval superstition, it was taken for granted that poisons could filter down through the earth and enter water supplies.

As the United States industrialized after the Civil War, pollution worsened and came to be seen as more than a merely local problem. By the dawn of the twentieth century, victims of pollution were seeking remedies at state and national levels. Polluting industries resisted fiercely, developing tactics that were to remain in use a hundred years later.

———

The "smoke problem" is centuries old. When firewood ran scarce in the England of the late middle ages, coal was used to fire lime kilns, forges, and other workshops. An investigative commission was established as early as 1285, and the offending trades were pursued with edicts and lawsuits. By Elizabethan times, coal was used to heat homes as well, and a dense winter mixture of smoke and fog came to afflict London.[5] The opacity of this murk and a characteristic yellow tint earned it the name of pea-soup fog. Its evanescent hues, changing by the hour, haunted artists and poets. These shifting colors were depicted in oil by the impressionists and painted in words by Oscar Wilde[6]:

> The Thames nocturne of blue and gold
> Changed to a harmony in grey;
> A barge with ochre-coloured hay
> Dropt from the wharf: and chill and cold
>
> The yellow fog came creeping down
> The bridges, till the houses' walls
> Seemed changed to shadows, and St. Paul's
> Loomed like a bubble o'er the town.

As coal burning spread to Europe and North America with the industrial revolution of the nineteenth century, the scourge of urban air pollution spread with it. Smoke abatement was a staple topic of municipal reform movements in the late nineteenth and early twentieth centuries, and many American cities, especially Midwestern industrial centers where the problem was most severe, adopted control ordinances. But these early ordinances often lacked real teeth, with industrial interests in particular able to secure escape hatches.[7]

Coal smoke, in the early years of industrialization, was treated as a local problem. Air pollution was made a national concern by the sulfurous emissions from copper smelting. The issue initially came to a head in Ducktown, Tennessee, where smoke damage extended across the state line into Georgia. The neighboring state sought relief from the U.S. Supreme Court, which in 1907 ordered the plant to reduce its emissions.[8] Lawsuits followed at smelters throughout the country.[9] Experts hired by plaintiffs—among them Robert Swain, who went on to become a leading chemist of the first half of the century—pioneered the scientific study of smelter smoke and its effects; they quickly demonstrated that the smoke contained the poisonous elements arsenic and lead along with sulfur.[10]

The smelter smoke issue reached a climax in President Theodore Roosevelt's struggle with the Anaconda Copper Company in Montana. After repeated death-dealing episodes of air pollution in the city of Butte, the company had relocated its smelter to a remote valley. But destruction of crops and livestock continued in the vicinity of the new plant. When Anaconda used its influence over compliant local courts to fend off a lawsuit filed by the aggrieved farmers, Roosevelt mobilized the power of the national government to seek a remedy. Ligon Johnson, the lawyer who had won the Ducktown case for the state of Georgia, was hired as the government's "smoke counsel." The president sought to reach a compromise rather than file suit, but the company rebuffed his entreaties, even after a meeting at the White House with Roosevelt and Attorney General Charles Bonaparte in the presence of three senators and the complaining neighbors.

In the absence of a federal air pollution statute, the government prepared to go to court over damage to the national forest, where trees were killed as far as 22 miles from the smelter. The lawsuit confronted daunting obstacles. The government faced the economic power of interlocking trusts; Anaconda

The Anaconda, Montana, copper smelter in 1903, five years before President Theodore Roosevelt initiated a pioneering environmental lawsuit to curb its air emissions. (Courtesy of the Montana Historical Society Research Center.)

was controlled by Standard Oil interests and could count on the support of Wall Street. Further, the government lacked scientific and engineering expertise and had no easy way to procure it. "The mouths of all the experts would be closed," Bonaparte told the president three days after their meeting with the company, "because [Anaconda] and the Standard Oil Company ultimately will reach them and control them wherever throughout the country and in whatever individual enterprises they are engaged..."[11]

The suit against Anaconda was a high priority for the president, who was kept informed by the attorney general of even such details as the search for expert witnesses. But only when Roosevelt was about to leave office was the government ready to go into court. The Taft administration, friendlier to business, waited 12 months before filing the suit. After another year a settlement was reached with Anaconda.

Under the settlement agreement, the company promised to minimize its emissions, with a three-member commission appointed to prescribe means of doing so. This commission launched studies of a wide variety of proposed control technologies under the auspices of the newly established Bureau of Mines. The investigations dragged on, and after a few years the government dropped its previous insistence on a cleanup and accepted the company's criterion for action—controls were required only if captured emissions could be sold to yield a profit. Research continued with diminishing energy and increasing subservience to Anaconda until the commission expired quietly in 1924.[12]

In the end, the only significant control measures were smoke-collecting devices—settling chambers, baghouses, and electrostatic precipitators—that captured arsenic but had little effect on sulfur emissions. Arsenic could be sold; market demand grew rapidly between 1918 and 1923 after it was shown that calcium arsenate could control the boll weevils that devastated cotton crops. In 1923, Anaconda added more precipitators; soon the smelter by itself produced nearly enough arsenic to saturate the pesticide market nationwide, and the company worked intensely to find additional uses. Spread across farm fields and sprayed on fruit, the arsenic captured in the precipitators created a new set of environmental problems that would be bequeathed to posterity.[13]

Subsequent administrations were not inclined to duplicate Roosevelt's legal activism against Anaconda, and Ligon Johnson left the Justice Department in 1914 to become smoke counsel for American Smelting and Refining Corporation at double his government salary. Although private parties continued to file suits through the 1920s, the smelter smoke issue largely faded from public view.[14] One controversy did gain prominence when the federal government was drawn in. This dispute concerned emissions from a zinc smelter in Trail, British Columbia, that damaged crops and timber across the border in the United States. The Trail case dragged on from 1927 to 1943 as the American and Canadian governments financed dueling teams of researchers. In the end, an arbitration panel assigned blame to Canada.[15]

———

Factories have contaminated streams since the dawn of the industrial revolution, and governments have long taken notice. As early as 1901, a survey article could report that the only jurisdictions in the United States that lacked statutes to control water pollution—which was understood to include both industrial waste and sewage—were Louisiana, Arkansas, Texas, South Carolina, and Indian Territory (the future Oklahoma). By 1924, all 48 states had laws that covered industrial waste. Both surface water and groundwater were always understood to be vulnerable to contamination; by the mid-nineteenth century, lawsuits over pollution of groundwater were numerous.[16]

Beginning in the 1890s, bills about water pollution were introduced regularly in Congress. The 1912 act that gave the Public Health Service its present name also authorized it to investigate the contamination of navigable streams. A team of engineers and scientists was assembled for this purpose on the site of a closed Marine Hospital in Cincinnati. They were joined there by the Public Health Service's new industrial hygiene team. By 1914, the Cincinnati laboratory had grown into an Office of Industrial Hygiene and Sanitation, which shortly thereafter was split into two divisions.[17] This small laboratory, working closely with the Bureau of Mines and sharing staff with it, began to create the governmental scientific competence whose lack had been felt so strongly when the smelter suits were undertaken. Over the course of six decades, this environmental staff would grow in size and scope, expanding into air pollution and radiation. Eventually, it would develop into two large government agencies, the Environmental Protection Agency and the National Institute of Occupational Safety and Health.

Industrial wastes were not the main concern of pollution fighters in the years around the turn of the twentieth century. Scientists had shifted their attention from chemical to biological causes of sickness as the germ theory of disease was elaborated by Pasteur, Koch, and others. They found much to worry them in the environment. Human wastes, previously poured into the ground in cesspools and privies, were now carried into streams by municipal sewers. Rivers were visibly fouled, and there was invisible damage too. Sewage discharges were identified in 1890 as the cause of urban typhoid epidemics along the Merrimack River in Massachusetts and the Mohawk River in New York. This discovery sharpened the focus on sewage, and state health agencies, soon joined by the federal Public Health Service, busied themselves with typhoid control and sewage research.[18]

Attention returned to industrial water pollution in the aftermath of the First World War. The rapid growth of petroleum production and the conversion of ships from coal to oil power brought forth a new pollution problem. Oil-fueled ships, needing ballast when fuel tanks emptied, filled the tanks with seawater. Upon refueling, whatever oil had been left in the tanks was dumped into the seas along with the water. More oily wastes entered the ocean from refineries and other plants along the shore. Birds and shellfish died, black tar disfigured beaches, and oil sometimes accumulated in amounts so great as to create a fire hazard. Owners of beach resorts in New Jersey and nearby states were first to raise the alarm; their National Coast Anti-Pollution League quickly gained backing from commercial fishermen, fire insurance companies, and other interests. In response, a national conference of fish commissioners was called by Secretary of Commerce Herbert Hoover in June 1921.

Competing solutions were quick to emerge within the government. On one side was the Army Corps of Engineers, which exercised the very limited federal powers over water pollution that had been created by the New York Harbor Act of 1888 and the Rivers and Harbors Act of 1899. It advocated strong legislation that would extend to inland waterways and cover pollution from mines and factories as well as ships. Hoover, on the other hand, came to the problem with a strong presumption for a constricted government role—one year later, he told the National Association of Manufacturers that the Commerce Department's regulatory powers were "a few inconsequential matters in connection with the safety of human life." Commerce proposed that only oil pollution from ships should be controlled, while other problems would be deferred for further study. The two agencies clashed over turf as well as policy; each asserted, with backing from friendly interest groups, a claim to run the new controls.[19]

Congressional hearings in the fall of 1921 brought forth a wide spectrum of competing interests. While manufacturing industries tried to narrow the scope of legislation, oil and shipping interests objected to being singled out for controls, and at the same time the various business groups joined in contending that lengthy study should precede action.[20] Although oil companies were the main actors, the chemical industry's trade association, the Manufacturing Chemists' Association, briefly entered the fray. Its secretary, John Tierney, introduced an argument for inaction to which his industry would turn again and again in future pollution debates:

> . . . in order to get out a measure that is scientific, that will do justice
> to all interests and injustice to none, and visit hardships as little as
> possible on those that will be affected, there should be some kind of

inquiry made...Without that information, I think you will inevitably be doing harm to many interests and injustices will occur.[21]

Information was indeed lacking, and Congress turned to the Bureau of Mines for an investigation of the problem. The newly formed American Petroleum Institute sprang into action and offered to fund the study. Under this arrangement, the head of the API's Division of Research, former Bureau of Mines director Van Manning, oversaw the Bureau's expenditures while representatives of API and the steamship owners helped to prepare the report. An API researcher also accompanied some plant inspections. Manning asserted a right of approval over the final report before its public release; the Bureau of Mines, whose policy to that point had been to insist on unfettered publication of work done under outside sponsorship, compromised by conceding Manning an informal review and approval.[22]

While the investigation proceeded, a strong oil control bill—whose scope did not, however, extend to other industrial wastes—passed the Senate in 1922. The bill died without a House vote at the end of the session, with charges flying that the speaker pro tempore had blocked it at the behest of the petroleum industry. During the new Congress, the Bureau of Mines issued its report. The study conclusions unsurprisingly mirrored earlier oil industry assertions, with a caution against singling out any one industry for controls and an assessment that land-based plants already had sufficient technology to prevent oil releases. The new Congress followed the report's recommendations, passing a bill that entirely exempted the refineries and other sources on the land, controlled vessels only in coastal waters and not in rivers, and failed to authorize funds for enforcement. To placate the resort owners and fishing interests, a follow-up study was authorized. This greatly weakened bill was signed into law by President Coolidge on June 7, 1924.[23]

The refiners' success in that legislative battle did not exorcise the specter of regulation. The Corps of Engineers in 1926 sent a detailed report on water pollution to Congress, recommending a substantial expansion of federal oversight. Self-regulation, the oil men responded, was more effective than government control, and their trade association tried to make good on that promise. The times were ripe for such an initiative; federal policy was guided by Hoover's vision of industrial self-government staving off the twin perils of excessive competition and central control. In 1927 the API conducted its own pollution survey, and it went on to compile a *Manual of Disposal of Refinery Wastes*, issued in three parts between 1931 and 1941.[24]

The committee that prepared the 1927 survey recommended that the trade association put teeth into self-regulation by setting up its own inspection and enforcement system. This proposal was quickly shot down by objections from Howard Pew of Sun Oil and other influential association members.[25] Compliance became the responsibility of the individual oil companies, setting a pattern for future efforts at industrial self-regulation. The API's decision not to establish a robust system of enforcement endowed with coercive powers was a turning point. It is impossible to say where the road not taken might have led. Would enforceable self-regulation have survived judicial scrutiny like the land-use planning and zoning ordinances that Hoover championed, or would it have been struck down like the industrial codes of the early New Deal? As things were, future efforts at environmental self-regulation would follow the pattern set by the API and have only the force of advice and exhortation.

Legislation aimed at controlling water pollution continued to be introduced in Congress over the ensuing decade. Only one bill, which would have extended the oil pollution act to inland waters, even made it to the hearing stage.[26] Until the New Deal arrived, impenetrable barriers stood in the way of any federal action on water pollution.

| The Rise of the Chemical Industry

There are laws aimed at pollution abatement, but they cannot bring clean air and water... it is invention and development, not legislation or regulation, that has proved our most reliable instrument of progress.

—Henry B. du Pont, 1952[1]

FOR MUCH OF ITS history, America depended on Europe for dyes, explosives, and finished chemical goods. Only in the twentieth century did the chemical industry become a central pillar of the economy and did its leaders gain commensurate political influence.

In England, Germany, France, and Switzerland, the Industrial Revolution was a chemical revolution as well. Demand from growing industries called forth large-scale chemical production while the advance of scientific knowledge in the continent's long-established universities engendered new manufacturing methods and then brought entirely new products into the marketplace. In the eighteenth century, acids were needed for the treatment of metals and the preparation of dyes, alkalis for making soap and glass, and chlorine chemicals for bleaching. Advances in organic chemistry in the early nineteenth century led by midcentury to the small-scale use of synthetic organic chemicals, synthetic dyes, explosives, and synthetic medicines. These developments were soon followed by large-scale manufacture of coal-tar dyes and modern explosives such as Nobel's dynamite. Pharmaceutical products such as antipyrin and aspirin, chlorinated solvents, and the plastic bake-lite followed at the turn of the century. By around 1900 most of the major branches of the globe-straddling chemical industry had been established.[2]

In the first half of the twentieth century, the chemical industry built on the progress of the previous century and emerged as a key sector of modern technology, with new synthetic products—plastics, fibers, coatings, pesti-cides, and pharmaceuticals—emerging to dominate markets. An early model of science-based chemical business was the dye industry that emerged in Germany between 1860 and the First World War. BASF, Bayer, and other firms worked with Germany's preeminent research universities to develop new products and manufacturing methods.[3]

America's chemical industry began soon after the Revolutionary War. At the urging of President Thomas Jefferson, who sought a domestic source of high-quality gunpowder, the immigrant French chemist Eleuthère Irénée du Pont de Nemours built a black-powder plant in Delaware in 1802. Irénée du Pont's company came to dominate America's powder business and achieved a near monopoly by the end of the nineteenth century. After reorganizing when a new generation of du Ponts took control in 1902 and consolidating its position in the explosives business, the company began to evolve into a chemical company.[4]

The chemical industry in the United States was at first slow to catch up with Europe. In 1850 a mere one thousand people were employed in the nation's 170 chemical plants. As western expansion and industrialization accelerated after the Civil War, the chemical industry began to grow rapidly; most of today's major American chemical companies got their start between the 1870s and the First World War.[5]

Many of the chemical companies, particularly the larger ones such as Monsanto and Dow, were founded by a single dominant figure. Herbert Dow began in 1889 to extract valuable chemicals from salty brines, often a waste product of oil wells, that lay deep below Ohio and Michigan.[6] John F. Queeny established Monsanto (named after his wife's family) in 1901; his first product was saccharin, the artificial sweetener discovered accidentally ten years earlier at Johns Hopkins University.[7] Other new firms established around the turn of the century would develop as well into large enterprises, including PPG, Olin Chemical, Union Carbide, Hooker Electrochemical, and American Cyanamid. The founders of DuPont, Monsanto, Dow, and other closely held firms passed control down to their children and their chil-dren's families—and with their companies they passed down strongly held political and philosophical convictions that continued to shape the culture of the industry decades after the founders' deaths.

The American chemical manufacturers depended on tariffs to shield them from the competition of the more efficient and technologically advanced overseas producers, and they were quick to organize. The Manufacturing Chemists' Association was established in 1872 by a group of sulfuric acid manufacturers, and it began lobbying in Washington just six years later. Flexible tactics were required in order to pursue the industry's narrow objective in an environment where educated opinion was hostile to "protectionism." By 1912, the association's politically adept leadership could report with entertaining cynicism how a tariff-cutting bill had been fended off:

> ...aided by a hopeless political situation in which all factions were endeavoring to accomplish nothing, but to lay the blame for such lack of accomplishment upon their political opponents who were respectively referred to as the friends of protected interests, the tariff agitation of the 62nd Congress has passed into history, and the chemical bill still slumbers.[8]

Propelled by the new continent's vast ore resources, nonferrous metals emerged as an early area of strength for American manufacturing. This is a domain where the chemical and metallurgical industries overlap; nonferrous metals such as aluminum, chromium, and zinc are often sold as chemical compounds rather than in metallic form, or are separated from their ores by processes more typical of chemical production than metal smelting. By 1914, the chemical industry employed 67,000 people. It was self-sufficient in many basic products and was the world leader in sulfuric acid. But it continued to import intermediate chemicals and lagged far behind Germany in the production of synthetic organic compounds.[9]

It was under the impulse of the First World War that the manufacturing of synthetic chemicals, already well established in Europe, began on a large scale in the United States. Even before the United States entered the conflict, the war transformed the chemical industry. Explosives makers like DuPont and Hercules ran plants to the limit to fill European orders, while a British blockade of German exports impelled the sudden launch into production of chemicals that had previously been imported. War-inflated profits were plowed into new plants, capital flooded in from investors, and by November 1917, the secretary of the American Chemical Society could say that the country's chemists had "accomplished within two years what it took Germany forty years to attain."[10]

Among the most significant technical innovations of the war years was a vast increase in the output and diversity of synthetic chemicals. Completely novel compounds without analogs in nature were created. Thus chlorine was introduced into benzene to yield the new moth killer paradichlorobenzene. Dyes and their intermediates, several of which were vital to the manufacture of modern explosives, were among the most important new products.

Expansion into new product lines was accelerated when the government seized German chemical patents as enemy property and granted licenses on favorable terms for their use by American firms. Once the war began, the government also directly financed the construction of plants. Seventeen companies began making dyes in 1917 alone, and by 1920 output of these chemicals was 15 times what it had been in 1914. Hooker Electrochemical, which made only bleach and caustic soda before the war, had 17 products by November 1918 and was the world's largest producer of monochlorobenzene.[11]

The Germans began in 1915 to use poison gas on the battlefield, hoping to use their superiority in the industry to break a stalemate on the front. The British and French of course responded in kind. The first poison gas was chlorine; soon after, both sides began developing new toxic agents such as mustard gas and chloropicrin.

The United States was slow to react to the advent of chemical warfare, and in 1917 American troops began to arrive at the front unprepared. A crash program to build plants to produce chemical agents was quickly launched, and the country's industrial might was applied on a scale that soon eclipsed the efforts of its allies. If the war had continued two months longer, the new government-built plants would have produced 200 tons per day of poison gas.[12]

After the war, the seizure of German patents was confirmed and made permanent by the Versailles peace treaty. The U.S. rights to the patents were held by the alien property custodian, Francis Garvan, who was a close associate of Attorney General Mitchell Palmer. Garvan sold 4,500 patents for $250,000, a small fraction of their value, to a nonprofit entity, the Chemical Foundation, that the Wilson administration had established under the control of the major American-owned chemical companies. Garvan then left the government and became president of the Chemical Foundation, a post he held for the rest of his life. The patents were made available under license to the American manufacturers, with the royalties used to support the industry's interests through lobbying, public relations, and the promotion of chemical education and research. The Republican Harding administration in 1922 filed a politically charged lawsuit challenging the transfer of patents, but the validity of Garvan's action was upheld by the courts.

Despite the advances made during the war, American know-how in chemical production still lagged far behind Europe. The German chemical companies were quick to regroup after the war and remained formidable competitors.

The American industry thus sought protection for new and old products and launched an intensive lobbying campaign. The wealth of the Chemical Foundation enabled it to spearhead the effort, while the Manufacturing Chemists' Association gained new influence when DuPont joined in 1919. This political work bore fruit in the 1922 Fordney-McCumber Tariff Act, which set especially high tariffs on the new synthetic organic chemicals. The Chemical Foundation would retain influence until its activities wound down after the expiration of the last of its patents in 1936 and Garvan's death in 1937.[13]

The first postwar years were hard; the problems of converting to peacetime production in a time of depressed business conditions were exacerbated by unfamiliarity with newly adopted technologies. But with the government behind it, the industry succeeded in overcoming these obstacles, and a wave of expansion and consolidation ensued. American Cyanamid, with strong Wall Street backing, and DuPont, flush with profits from wartime munitions sales, gobbled up smaller family-owned chemical firms.[14] DuPont's financial strength was magnified by the growth of General Motors, in which it had acquired a controlling interest in 1918. DuPont in 1921 introduced a new management structure that placed autonomous operating divisions under centralized financial controls. A similar arrangement brought order to the chaotic operations of General Motors. The organizational model developed by DuPont and General Motors was hailed for decades afterward as an exemplar of good management.[15]

———

The chemical industry's response to steadily increasing demands for control of its air and water emissions would be heavily colored by its political orientation. The men who led major chemical companies adhered to an ideology of laissez-faire.[16] Their denunciations of government interference with business would earn them a prominent position on the far right of the political spectrum. Yet, in practice, the chemical manufacturers objected to government intervention only when it escaped their own control. The government had played a dominant role in fostering the industry's growth through munitions orders, tariff protection, paying for construction of plants, and the seizure of German patents.

The contretemps over the Chemical Foundation unavoidably influenced the industry's partisan allegiances, and during the 1920s chemical companies veered toward the Democrats. The partisan divide over prohibition of alcoholic beverages reinforced this trend—the precedent of banning a harmful chemical compound did not appeal to the makers of substances that were unarguably more poisonous. DuPont, the leading firm within the industry, also took the lead on the national political stage. The du Ponts had been conservative Republicans; Coleman du Pont, a long-time member of the Republican National Committee, was elected to the Senate from Delaware, and the entire family backed Warren Harding in the 1920 election. But the executives of the chemical company were soon to become major backers of the Democratic Party. The change of parties was justified with the contention that the Coolidge and Hoover administrations were insufficiently pro-business. The du Ponts became leading "wets" as well, a stance that harmonized with the family's ingrained hostility to government control of business; a further appealing thought was that a tax on alcohol might raise enough revenue to repeal the income tax.

After Coleman du Pont fell ill and resigned from the Senate in 1928, John Raskob emerged as the most prominent DuPont man in politics. Raskob, inclined toward the Democratic Party by his strong Catholic religious affiliations, had been hired as a bookkeeper by Pierre du Pont in 1900 and rose to be treasurer of DuPont and to chair the Finance Committee of DuPont-controlled General Motors. When the Democrats nominated the "wet" Al Smith for president in 1928, Raskob left the auto company to serve as chairman of the Democratic National Committee. For several years after Smith's defeat, Raskob personally financed most of the national party's expenditures. Upon his return in 1930 to General Motors, a close associate, Jouett Shouse, took charge of the day-to-day running of the party.[17]

In the years that followed the 1932 election, these deeply conservative men recoiled strongly against the activism of Franklin Roosevelt's New Deal. Their vehicle was the American Liberty League, formed to promote the principle of government noninterference in the economy. Initially billed as a nonpartisan association, it soon emerged as the voice of the New Deal's most uncompromising opponents. The Liberty League was conceived in a discussion among DuPont directors at the conclusion of a company board meeting in late 1933 and established with financing from Irénée du Pont, the vice chairman of the board; his brother Lammot, the president; and their family. Al Smith, Jouett Shouse, and 1924 Democratic presidential candidate John W. Davis at first gave the organization a nonpartisan veneer, but it quickly became the center of resistance to Roosevelt's reforms.

The Liberty League framed its program as a defense of constitutional liberties. The New Deal was condemned without nuance as not merely unconstitutional, but a stepping stone to dictatorship. The only remedy for economic troubles was to let things be; if Roosevelt had not interfered, the depression would have resolved itself. The Jeffersonian rhetoric fell flat; the du Ponts' avid pursuit of monopoly economic power was known to all. The League was left to stand on a platform of resistance to New Deal economics. Unvarnished laissez-faire had little appeal on its own in the aftermath of the 1929 crash, so the du Ponts sought allies where they could; they were not above financing openly racist opposition to Roosevelt among southern Democrats.[18]

Other politically prominent chemical manufacturers shared the du Ponts' hostility to government interference with business. One was Elon Hooker, president of the Manufacturing Chemists' Association from 1923 to 1926. Hooker, trained as a civil engineer, first came to prominence by reorganizing New York's state-run Erie Canal for Theodore Roosevelt, whose election as governor in 1898 had been propelled by scandals in the canal's administration. Leaving government service after Roosevelt became vice president in 1901, he established a business to exploit the electric power potential of Niagara Falls. This was Hooker Electrochemical—the company whose later dumping would lead to the Love Canal disaster of the 1970s. Hooker's close friendship with Roosevelt, forged during his state service, gave him entrée into national politics. He served in the 1912 presidential campaign as national treasurer of Roosevelt's Progressive Party and unsuccessfully sought the Republican nomination for governor of New York in 1920.[19]

The object of the industrialist's admiration and loyalty was not Roosevelt the trustbuster and conservationist but the hero of San Juan Hill, the man of the big stick and 100% Americanism. Following his failed campaign for governor, Hooker took the reins of a right-wing advocacy group called the American Defense Society. Originally established in 1915 to promote U.S. entry into the First World War, the society's postwar program mixed promotion of patriotism and military preparedness with assaults on progressives of all stripes as communist-inspired. In the months before the 1924 election, when independent candidate Robert LaFollette, running with the support of labor and non-Communist leftists, won 17% of the vote against pro-business candidates of both major parties, the society issued anti-LaFollette pamphlets by the tens of thousands. They bore such titles as "The Revolution Against the American Government" and "LaFollette-Socialism-Communism."[20]

Another politically active chemical executive was William Bell of American Cyanamid, president of the Manufacturing Chemists' Association from 1933 to 1936. During the early years of the Franklin Roosevelt administration, Bell led the trade association in lobbying against any form of government control even as it benefited from high tariffs that curbed foreign competition. He denounced New Deal planning with comparisons to the Soviet Union and promised that recovery would come faster if government would just leave business alone. When Roosevelt's budget director, former Congressman Lewis Douglas, broke with the New Deal in 1934, Bell gave him a well-paid position as a company vice president. Lacking industry experience, Douglas served as a "trouble-shooter and economic analyst" while continuing to pursue his political ambitions.[21]

Partisan politicking by chemical manufacturers reached a peak when Roosevelt was challenged by Alf Landon in the 1936 election. Direct chemical industry donations to the candidates, excluding expenditures through the Liberty League, amounted to $160,000 for Landon and only $1,000 for Roosevelt. William Bell was the chief Republican fundraiser in the run-up to the party's 1936 convention. DuPont contributions, given directly and through the Liberty League, provided almost 8% of the Landon campaign's total resources—$645,000 out of $8.4 million. "Without Liberty League money," Republican national chairman John Hamilton later confessed, "we couldn't have had a national headquarters."

The Liberty League was an easy target for New Dealers, who had little difficulty portraying it as a millionaires' club. James Farley, Raskob's successor as the Democrats' national chairman, joked that the Liberty League should be called the American Cellophane League. "First, it's a du Pont product and second, you can see right through it." The du Ponts reinforced this message with their own actions: the courts ruled that Pierre du Pont and John Raskob had evaded two million dollars in income taxes through sham stock transactions after the crash. At the Democratic national convention, Farley took directly after the family with this verbal assault:

> Behind the Republican ticket is the crew of the du Pont Liberty League and their allies, which have so far financed every undercover agency that has disgraced American politics with their appeals to race prejudice, religious intolerance and personalities so gross that they had to be repudiated by the regular Republican organization.

The convention culminated in Roosevelt's famous acceptance speech, calling a generation to a "rendezvous with destiny" in the battle against economic royalists.[22]

Election Day brought a landslide victory for Roosevelt and the New Deal. This stunning defeat so discredited the Liberty League that its public activities were quickly wound down. The organization lingered quietly on, financed entirely by the du Ponts, until finally ceasing operations in 1940.[23]

While the du Ponts withdrew from the political foreground, fearing the consequences of public opprobrium for their company's sales,[24] another chemical mogul emerged as an outspoken critic of the New Deal. Edgar Monsanto Queeny, a world-trekking adventurer and documentary filmmaker, was heir to the chemical company and chairman of its board of directors. The Monsanto interests had made heavy contributions to the Landon campaign alongside DuPont and American Cyanamid, and Queeny was a director of the National Association of Manufacturers and the National Industrial Conference. Queeny's 1943 book *Spirit of Enterprise* did not merely condemn the political and philosophical foundations of the New Deal, it attacked the intelligence and patriotism of its architects. Queeny echoed his chemical industry predecessors in arguing that business should be left to conduct its affairs while government offered help and advice. The citizenry might in consequence suffer some degree of insecurity and financial sacrifice, he conceded—but centralized planning and control are contrary to an essential American trait of individualism.[25]

———

With the wind of technological progress behind its back, the chemical industry flourished in the 1920s and 1930s. New synthetic chemicals appeared in large numbers. Artificial fibers—rayon and nylon—and plastics entered the marketplace, along with a wide variety of solvents and other industrial chemicals. Coal tar, from which the first synthetics were made, was displaced by the by-products of cracking processes used in petroleum refining, creating what came to be known as the petrochemical industry. By 1939, petroleum and natural gas had largely supplanted coal tar as raw materials for chemical synthesis, and oil giants like Standard of New Jersey and Shell became major participants in the chemical business.[26]

The chemical manufacturers weathered the crash of the early 1930s with relatively little difficulty. With their profit margins protected by tariffs, which were raised to new heights by the Smoot-Hawley Act of 1931, and by governmental tolerance of informal arrangements that suppressed price competition, they were among the most profitable of the nation's businesses. The industry's new prominence was marked on the skyline of New York, the country's business capital. When Rockefeller Center opened in 1934, American Cyanamid moved its headquarters to the top floors of the 70-story RCA Building. Pierre du Pont and John Raskob financed the construction of the Empire State Building, the world's tallest structure, raising its height to 102 stories to outdo their automobile rival's Chrysler Building.[27]

The Second World War brought another burst of expansion. New industries like synthetic rubber and organic pesticides were built from scratch; output of existing products grew rapidly, spurred by supply contracts and the financing of new plants; and the pace of technological advance was sharply accelerated by central planning of research and mandated sharing of proprietary knowledge.[28] The return of peace found the industry more prosperous than ever. DuPont remained the industry leader, with sales of $783 million in 1947, followed by Union Carbide at $522 million, Allied Chemical & Dye at $365 million, and American Cyanamid at $215 million.[29]

The leaders of DuPont—two brothers, a brother-in-law, a nephew, and a cousin—at the New York World's Fair of 1939. They are previewing the company's Wonder World of Chemistry pavilion, where nylon was first introduced to the public. From left, Irénée du Pont, vice chairman of the board; Henry B. du Pont, later a company vice president; Ruly Carpenter; William du Pont Jr., director; and Lammot du Pont, president. (Courtesy of the New York World's Fair 1939–1940 records, Manuscripts and Archives Division, The New York Public Library, Astor, Lenox and Tilden Foundations.)

The du Ponts continued to keep a low public profile, their reticence reinforced by a series of antitrust suits against the industry that culminated in a successful attack on their control of General Motors. But the family's opinions on economic and political questions were unaltered and unforgotten. Donations flowed to a panoply of far-right educational and lobbying groups. When the steel industry's hard line in 1946 wage negotiations led to a strike, Alfred Friendly of the *Washington Post* saw a coordinated assault on labor, led by a du Pont-Sloan-GM-Grace-Bethlehem coalition.[30]

As the effects of pollution grew more severe with wartime and postwar industrial expansion, the practical measures required to bring them under control came more frequently to run afoul of the chemical manufacturers' political and economic views. The reality of the emissions problem, known to the industry even before the war, was now clear to all. The industry's quest for profits had overridden its hostility to government intervention when it sought protection against foreign competition through tariffs. Would it make a similar exception for the environment, allowing government to act as referee so that polluters gained no economic advantage? Or would it insist that the market rule, with corporations left to their own devices?

Henry B. du Pont, a company vice president, met this question head-on in an address to the Second National Air Pollution Symposium in 1952. His response showed that the du Ponts still adhered to the faith of the Liberty League. "There are laws," he argued, "aimed at pollution abatement, but they cannot bring clean air and water..." Laws were only effective when the control agencies cooperated with industry. Du Pont concluded with an appeal to history:

> It seems clear to me that our greatest promise in abating pollution lies in giving full reign to advancing technology. As Americans have found in every field, it is invention and development, not legislation or regulation, that has proved our most reliable instrument of progress. The farm reaper was not invented because of legislation or land reform, yet it had a more profound effect upon agriculture than any law, before or since. Long before the Emancipation Proclamation, the work of inventors like Whitney and Howe had doomed human slavery. Child labor was not abolished just by statute, but by productive machinery that made such work unnecessary. I think it can be well demonstrated that all of our social gains have had a similar history; modern technology is the greatest reformer of modern times.[31]

A man who thought free enterprise deserved more credit than Lincoln for freeing the slaves would never entrust government with the responsibility for cleaning up pollution. The chemical industry would do what it would do on its own. When outsiders sought to make it do more, it would resist.

| Royd Sayers' Service Bureau

*{Secretary of Commerce Herbert} Hoover said that the Bureau of Mines was cre-
ated as a service bureau for the mining industry and that it is his purpose to develop
its activities.*

—*New York Times*, June 5, 1925

E ARLY IN 1917 TWO young Public Health Service doctors arrived in
Butte, Montana, and prepared to begin a study of the lung diseases and
other ailments that afflicted the city's copper miners. Here came together
for the first time a pair of men whose scientific and political influence would
shape environmental policy for decades to come.

Miners' health had been a concern of the federal government for all of seven
years, since the creation of the Bureau of Mines. The two doctors were on loan
to that agency, among whose purposes it was to ameliorate the health and
environmental effects of mining.[1] The Bureau was now relocating its medical
research to the city that already housed the smelter smoke investigation.

One of the physicians, Royd Sayers (1885–1965), was new to industrial
medicine. Born in the small Indiana town of Crothersville, where his father
owned a jewelry store, Sayers had studied chemistry at the University of
Indiana. After working as a chemist for three years, he enrolled in the medical school of the University of Buffalo, teaching chemistry while pursuing
his studies. Upon completing his internship in 1915, he joined the Public

Health Service. Sayers was first assigned to work on infectious diseases, but soon the military and industrial needs of the European war induced a rapid expansion of the PHS's efforts in industrial hygiene. Sayers' chemistry background made him a natural choice for this work, and in early 1917 he was detailed to the Bureau of Mines.[2]

Sayers' companion was his superior, Anthony Lanza (1884–1964), whose impressive title of chief surgeon of the Bureau of Mines placed him at the head of a medical team that could be counted on one's fingers. When Lanza joined the Public Health Service in 1907, he had been assigned to a tuberculosis sanatorium, where he gained expertise in lung diseases. That knowledge earned him a detail in 1914 to the new Bureau of Mines to direct a program of research on the health of zinc and lead miners in Joplin, Missouri. Silicosis, a lung disease caused by breathing rock dust, was spreading as mechanization made mines dustier; Lanza was to make his reputation by replacing the ill-defined concept of miners' consumption with the disease of silicosis, whose specific cause was silica dust. Sayers would collaborate with Lanza on silicosis research for decades to come.[3]

The Bureau's work on miners' health was soon interrupted by the United States' entry into war and the need for an expanded industrial hygiene program to support war production. In 1917 Lanza went back to the Public Health Service as chief of the Industrial Hygiene Division. He held that position until he left government employment in 1920. After helping to organize Australia's industrial hygiene program and serving as executive officer of the National Health Council, a public health umbrella group, he was hired in 1926 by the Metropolitan Life Insurance Company as its assistant medical director. Ostensibly, Lanza's job at Metropolitan was to provide industrial medicine and hygiene services to companies for which the insurance company had written group life policies. Actually, much of his time was devoted to helping company attorneys defend against workers' compensation claims, and in this capacity he worked with lawyers for numerous industries afflicted by hygiene problems. In 1947, nearing retirement age at Metropolitan, he joined the faculty of the New York University School of Medicine as the head of a new Institute of Occupational Medicine, whose activities were planned under the guidance of a committee composed of corporate medical directors and other executives.[4]

Sayers, who remained a uniformed Public Health Service officer, spent the next 16 years on loan to the Bureau of Mines. In 1917 he took Lanza's place as chief surgeon of that agency, and a few years later he added the title of chief of the Health and Safety Section. Sayers was a tireless worker who rarely took vacations; when he traveled overseas on behalf of the mining agency, much of his sightseeing was done underground. The scope of his work extended far beyond the extractive industries that were the bureau's main concern; he designed the ventilation system for the Holland Tunnel, the pioneering automotive connection beneath the Hudson River into New York City, and developed helium-oxygen atmospheres for deep underwater diving. In 1933 he followed in Lanza's footsteps and returned to the Public Health Service as chief of the Industrial Hygiene Division. There the pace of his scientific and administrative work grew yet more frenetic; by 1939 he led a staff of 90. He was an inveterate joiner, and the list of his committee memberships fills half a page. The apex of his career was reached in 1940 with appointment as director of the Bureau of Mines, where he remained for seven years. After retiring from government service in 1948, he consulted on workplace health problems until falling ill in 1963.[5]

As he toiled, Sayers moved far from his Indiana origins. In 1923, he married Edna Linnen, who he had first met at Anthony Lanza's home in Butte. Edna was the daughter of Interior Department inspector Edward Linnen, a man unfazed by political controversy and unafraid of making enemies. Edward Linnen had escaped firing through the protection of the secretary of the interior after his report accused a Senate ally of Theodore Roosevelt of misappropriating public land; the president wrote that the evidence "deeply discredits the worth and judgment of Inspector Linnen." In another high-profile investigation, Linnen drove the famed football coach "Pop" Warner out of his position at Carlisle Indian School by pursuing him over a series of minor transgressions, one of them the use of foul language in the locker room.[6]

Edna craved the social acceptance disdained by her father, often absent from their modest Minneapolis home, and her social climbing placed a heavy burden on the family budget. In 1932 the Sayers purchased the house where Robert E. Lee had been raised in Alexandria, Virginia. The Confederate military commander's boyhood home was a 6,000-square-foot structure on a half-acre lot in the historic Old Town. The new owners immediately began extensive restoration of this building, which had connections going back before Lee's birth to when George Washington was a regular dinner guest. The Sayers' entertaining in the lavishly furnished residence made its way onto the *Washington Post*'s society page.[7]

In purchasing their new home, the Sayers—neither of whom appears to have had inherited wealth[8]—must have made a large payment not recorded in the deed. According to that document, the house was bought with a ten-dollar down payment and $14,000 in seller financing. But the property surely was worth more than $14,000. It had sold for $12,000 plus ground rent in 1799, and the sellers' note required the Sayers to carry $15,000 in fire insurance, coverage that would have been unobtainable if the combined value of the house

Dr. Royd Sayers taking a blood sample from a woman, likely his wife
Edna, at about the time of the Sayers' wedding in 1923. As chief
surgeon of the Bureau of Mines from 1917 to 1933 and director of the
Industrial Hygiene Division from 1933 to 1940, Sayers downplayed the
dangers of leaded gasoline, coal dust, and pesticides. (Courtesy of the
Library of Congress, National Photo Company collection.)

and the half-acre lot was a thousand dollars less than the amount of the policy.
Nor is the tiny down payment explicable by market conditions. The depression
economy of 1932 created a buyers' market for real estate, but the Robert E. Lee
house is a Virginia landmark that would have attracted bidders in any year.[9]

———

As the manufacturing of new synthetic compounds expanded rapidly during
the 1920s, health problems started to emerge. Chemicals that would later be
major contaminants of the environment were seen first to afflict the workers
and consumers who were exposed at the highest levels. Testing was needed,
and the government now had the facilities and personnel it had lacked during
the Teddy Roosevelt administration.

The introduction of leaded gasoline was a watershed in the mass production of synthetic chemicals. The fuel was developed in the early 1920s by
General Motors, which sought to outclass Ford's mass-produced cars with
better performance and more varied design. Powerful engines demanded
higher-octane fuel to prevent "engine knock," and for this purpose GM's
Thomas Midgley came up with tetraethyl lead. When this compound was
added to gasoline in small amounts, the knock disappeared as if by magic.

Other formulations of high-octane fuel were known, but all had their problems. Ethyl alcohol, called ethanol by chemists, could solve the knock problem, but the new prohibition law made it hard to get supplies for experiments. Moreover, Midgley's boss, the famed automobile engineer Charles Kettering, feared that production of ethanol in the quantities required to fuel cars would use up food crops. Tetraethyl lead promised enormous profits, so General Motors pushed ahead. To manufacture the additive, the auto company contracted with its sister company DuPont and with Standard Oil of New Jersey. Production lines were opened at DuPont's Deepwater plant in southern New Jersey and Standard's Bayway refinery in Elizabeth.[10]

Questions were quickly raised about whether lead, long known as a poison, was safe to use in fuel. William Mansfield Clark, the head of the Chemistry Division in what would soon become the National Institute of Health, wrote to the Assistant Surgeon General in October 1922 to warn of "a serious menace to the public health." Clark was a distinguished scientist who had pioneered the measurement and control of pH in biochemistry. The Public Health Service's response to this warning was that research was needed, but experiments would take time. It was decided to contact General Motors first.[11]

The inquiry into leaded gasoline began with a letter from Surgeon General Hugh Cummings to Pierre du Pont, chairman of both GM and DuPont. A response came from Midgley, assuring Cummings that there was little hazard but conceding that there had been no experiments. General Motors offered to pay for a study by the Bureau of Mines. The Ethyl Corporation, the joint manufacturing venture that Standard Oil and DuPont had by now established, was granted approval over publication of the results. Sayers, as the Bureau's chief surgeon, was in charge.[12]

The dangers of the new leaded gasoline came forcibly to public attention in October 1924. Five workers at the Standard Oil plant in Elizabeth died and 35 others were poisoned by tetraethyl lead. With the victims suffering hallucinations and paranoia, "loony gas" made front-page headlines in New York newspapers. News of deaths at DuPont and GM soon emerged, and the *New York Times* uncovered more than 300 previously unreported cases of lead poisoning at the DuPont plant in Deepwater.[13]

The Bureau of Mines' preliminary report was issued one day after the fifth death in Elizabeth. Its conclusion that tetraethyl lead was not an environmental toxin came under sharp attack from public health advocates and independent scientists. Sayers and Emery Hayhurst of the Ohio Department of Health fought back, using their positions as public employees to defend the safety of leaded gasoline in professional forums and before the public. On Hayhurst's part, at least, this stance was not altogether disinterested. A letter he wrote to Sayers has come to light in which Hayhurst identified himself as "Consultant to Ethyl Gasoline Corporation."[14]

Whatever Sayers' personal motivations may have been, support for leaded gasoline was undoubtedly what his superiors at the Bureau of Mines now wanted to hear. In 1925, the agency was transferred from the Interior to the Commerce Department. This move signaled an abandonment of the bureau's progressive-era roots as a health and safety agency; Commerce Secretary Herbert Hoover made no bones about his orientation and told a press conference that the Bureau of Mines was a service bureau for the mining industry. The bureau now sold its services as a consultant to private corporations, calling when needed on the resources of other government agencies such as the Public Health Service. Reversing the earlier policy of open publication, research contracts routinely gave sponsor companies veto power over release of results. In at least one case, behind-the-scenes assistance was provided to corporations defending lawsuits.[15]

With newspaper publicity continuing, and local governments imposing bans on leaded gasoline, Surgeon General Cummings in May 1925 convened a national conference. After a spirited debate, the Ethyl Corporation announced it was suspending production of the fuel pending study by a committee of university scientists selected by the surgeon general. The committee commissioned a field investigation by the Bureau of Mines, with Sayers in charge.

The study population consisted of 77 chauffeurs and 21 filling station attendants from Dayton and Cincinnati who used leaded gas. Each underwent physical exams and gave blood and stool samples. Control groups that handled unleaded gasoline were similarly analyzed. Many in the group that worked with leaded gas had visible damage to their red blood cells, and elevated levels of blood lead were measured in some, but no clinical symptoms of lead poisoning were observed.

Study results were presented to the surgeon general's committee in draft form in December 1925, and after much internal debate the committee issued its conclusions the following month. It recommended that leaded gas production resume while studies continued with government funding. Questions about the integrity of the study have been raised by historians Mark Neuzil and William Kovarik. The fuel used by the study population may not have been ordinary leaded gasoline. The expert committee was told by James Leake, a Public Health Service epidemiologist who led the field work, that gasoline samples from service stations near the study group's workplace contained less than half as much lead as expected. But this information was omitted when a final written report was published two years later.[16]

The committee's recommendation that leaded gasoline go back on the market was rapidly adopted, but its request for further government studies would not be fulfilled for 40 years. Until the 1960s, the only studies of the use of tetraethyl lead were funded by the industry and carried out by Robert Kehoe (1893–1992), a professor at the University of Cincinnati. Kehoe served concurrently as the Ethyl Corporation's medical director, and he was a ferocious partisan of the now discredited concept that there is a safe level of exposure to lead.

Robert Kehoe was to exert wide influence in industrial health. He had joined the faculty of the University of Cincinnati immediately after his graduation in 1920 from the university's medical school, and he remained there for his entire career. When workers making tetraethyl lead at Midgley's pilot plant fell sick and died in April 1924, he was brought in to determine the cause, and he helped frame industry strategy throughout the subsequent controversy. At the 1925 national conference he propounded a position which synthetic chemical manufacturers would adopt in many subsequent controversies. Tetraethyl lead should be banned if an "actual hazard" was demonstrated—and only then. In the face of scientific uncertainty, society should err on the side of utility; demonstrable economic benefits should always outweigh unproven risks. Chemicals, in short, were innocent until proven guilty.[17]

The research that might discover an actual hazard from tetraethyl lead was in Kehoe's hands. With money from GM, DuPont, and the Ethyl Corporation, he established the Kettering Laboratory in 1930 to carry out contract research on hygiene problems for industry. Kehoe, who was paid only one dollar a year by the university and took his salary from his industrial sponsors, willingly granted the sponsors a veto over disclosure of research results. This policy gave Kettering a substantial fundraising advantage over the industrial hygiene laboratory at Harvard Medical School, where scientists expected to publish their findings. Soon the Kettering Laboratory displaced Harvard as the nation's preeminent center of industrial hygiene research. Kehoe held leading positions in numerous scientific associations involved with occupational health, and his laboratory was active as well in training students in the field. Many of his students did not share his biases; among them was Eula Bingham, who became director of the Occupational Safety and Health Administration in 1976 and was the most energetic leader in the agency's history.[18]

Leaded gasoline was surely on the mind of Thomas Midgley when, in 1928, he was asked to develop a new refrigerant for General Motors' Frigidaire division. Refrigerator sales had been zooming since the mid-1920s, but the refrigerants then in common use—ammonia, sulfur dioxide, and chloromethane—were all toxic. Chloromethane was especially dangerous because

Dr. Anthony Lanza, chief of the Industrial Hygiene Division in 1917–18 and associate medical director of Metropolitan Life Insurance Company from 1926 to 1947. At Met Life and later, he oversaw research on the hazards of silica dust, asbestos, and chromium on behalf of manufacturing companies. (Courtesy of the National Library of Medicine, History of Medicine Division.)

Dr. Robert Kehoe of the University of Cincinnati, who directed confidential research for chemical manufacturers at his Kettering Laboratory. Sponsors included the makers of leaded gasoline and the factories that caused the Donora air pollution disaster of 1949. (Courtesy of the National Library of Medicine, History of Medicine Division.)

its lack of odor meant that leaking refrigerators would not give warning to those breathing the fumes.

Within days of starting his work, Midgley identified short-chain carbon compounds containing both chlorine and fluorine—chlorofluorocarbons, or CFCs—as promising candidates. He then turned to the same scientists who had worked on tetraethyl lead. After initial toxicity tests by Kehoe gave promising results, intensive animal tests were carried out by Sayers and William Yant, a coauthor of the leaded gasoline study, at the Bureau of Mines. Dogs, monkeys, and guinea pigs were made to breathe high concentrations of dichlorofluoromethane 7 to 8 hours a day for 83 days.[19]

This work gained in urgency in July 1929, after a rash of sudden deaths in Chicago apartments. A coroner's jury led by Morris Fishbein, editor of the *Journal of the American Medical Association,* ascribed the death of 28-year-old Violet Clark, along with 14 earlier fatalities, to leakage of chloromethane from refrigerators. Although the evidence was strong—the compound was detected at high concentrations in Mrs. Clark's apartment, and two guinea pigs left in cages in front of her refrigerator were found dead when the jury reentered the apartment after 30 hours—refrigerator and chemical manufacturers questioned the jury's finding. A ban on chloromethane refrigerators issued by the city boiler inspector was quickly overruled by a city council committee, which called instead for further research by the city Health Department before action was taken. Federal agencies soon chimed in with reassurance. A joint statement by the Public Health Service, Bureau of Mines, and Bureau of Standards aimed at relieving "undue anxiety in the mind of those possessing household refrigerating systems." The hazard of chloromethane, the agencies said, was small, and it would soon be eliminated by adding odors or finding a new refrigerant.[20]

Once the testing was successfully completed, a CFC dubbed Freon entered the refrigerant market in December 1930. It was manufactured by a joint venture of DuPont, which earlier in the year had acquired a leading maker of chloromethane, and GM. The product was an immediate success and proved highly profitable.[21]

Tetraethyl lead and chloromethane had proved hazardous soon after they entered the consumer market, and both times the chemical industry's initial reaction was to study and stall. Efforts to have the products taken off the market were rebuffed, and the same team was called in for both studies. Events then took very different courses. Study of tetraethyl lead became an excuse for decades of inaction. Study of refrigerants led quickly to substitution of the much safer CFCs for chloromethane. Four decades later, these compounds would be found to endanger the ozone layer of the stratosphere, but Midgley and his associates can hardly be faulted for failing to anticipate this—in 1930 the chemical reactions by which stratospheric ozone breaks down into ordinary oxygen had just been identified, it was after the Second World War that other chemicals in the air were found to accelerate these reactions, and only in 1974 was the importance of chlorofluorocarbons recognized.[22]

Although the safety debates that led to acceptance of the octane booster and the new refrigerant took different courses, the outcomes of the two stories have a common element. DuPont emerged at the end of each with a proprietary product that dominated the market and yielded handsome profits.

As it pursued studies of new chemicals, the Bureau of Mines also expanded its research on miners' diseases into the coal mines. By 1924 it became clear that the illnesses of coal miners were something more than the silicosis suffered by metal miners; coal dust itself must cause lung disease. In February of that year, Daniel Harrington, a Bureau safety engineer who had worked with Sayers and Lanza in Butte, gave a stern warning to the American Institute of Mining and Metallurgical Engineers:

> It is entirely probable that a much greater number of men who have worked in our coal mines die annually of bronchitis, pneumonia, miners' asthma or other diseases caused directly or indirectly from coal dust, than die from mine explosions.

Harrington made similar points in *Coal Age* and the *Journal of Industrial Hygiene,* adding an appeal to company doctors to reveal what they knew to public health authorities. But objections from coal companies led the Bureau of Mines to prevent publication of the details of the studies. Royd Sayers contradicted Harrington and offered support for the mine owners' view of the science. Coal dust, he contended, is nearly harmless; only silica particles cause lung disease.[23]

In the waning days of the Hoover administration, Sayers returned to the Public Health Service as chief of the Industrial Hygiene Division. With him came the Bureau of Mines' testing laboratory. This transfer gave Sayers a three-year reprieve before coming under the supervision of New Deal appointees; Surgeon General Cummings' fixed term would not expire until 1936.[24]

With Sayers newly arrived at its helm, the Industrial Hygiene Division in April 1933 undertook an investigation of coal-mining diseases for the state of Pennsylvania. The study grew out of a legislative debate over whether to bring these illnesses under its workers' compensation law. Sayers chose to focus the investigation on a study group composed entirely

Royd Sayers in 1932. (Courtesy of the National Library of Medicine, History of Medicine Division.)

of active miners, relegating ex-miners to a separate, smaller study. This research design had the effect of understating the prevalence of lung disease, since anyone who had stopped working as a result of illness was excluded from the statistics. The high incidence of lung disease among coal miners still could not be denied, but the study report marginalized the coal miners' problem by calling the condition it observed "anthracosilicosis" and failing to analyze the relationship between coal dust and disease. The view that silica was the unique cause of miners' lung diseases was left unchallenged.[25]

A disaster in West Virginia would soon bring the problem of industrial illness much greater public prominence. In 1930 and 1931, Union Carbide had built a series of tunnels near the town of Gauley Bridge to divert the flow of the New River past a dam. The waters powered a factory that made chemical additives for the steel industry. The tunnels were excavated hurriedly through high-silica sandstone with little effort at dust control, using a labor force composed mostly of inexperienced Black migrants from the South. The result was an epidemic of acute silicosis; a later epidemiological study estimates more than 700 deaths among the few thousand workers employed in the tunnels.[26]

The hundreds of lawsuits filed by victims of the Gauley Bridge tunnels were only a small part of a flood of litigation filed in the early '30s by employees claiming exposure to silica dust; by early 1935, the aggregate amount of the claims was estimated as "well in excess of $100 million." Employers' previous approach to the dust problem was no longer viable once the New Deal threatened their cozy relationship with the government. Daniel Harrington had replaced Royd Sayers as head of the Health and Safety Section of the Bureau of Mines, and in 1934 the bureau itself was transferred to the Interior Department, whose secretary was the committed progressive Harold Ickes. Corporate defendants could still count the Industrial Hygiene Division as a friend, but they could not rely on it to come automatically to their aid. And the Labor Department, where deference to business interests was entirely lacking, was now seeking to add health problems to its agenda. A new strategy was needed, with new institutions to carry it out.[27]

Something surely had to be done about silicosis, but industry wanted corrective action on its own terms and under its own control. Thus was created the Air Hygiene Foundation, to be renamed a few years later the Industrial Hygiene Foundation. This organization, sponsored by a coalition of major corporations, was housed at the Mellon Institute of the University of Pittsburgh. Its formation was first proposed in September 1934, one month after Interior Secretary Ickes succeeded in overcoming opposition to his selection of a nonpolitical engineering expert to head the Bureau of Mines. The initiative came from the sand, glass, and refractory industries, but interest was much broader. A first meeting in Pittsburgh in January 1935, billed as a "Symposium on Dust Problems," attracted more than 200 executives. A smaller committee followed up, developing a comprehensive program that covered research, exchange of technical information, promulgation of standards, defense against lawsuits, and lobbying.[28]

The aim of the new foundation seems to have been to replicate the functions that the Bureau of Mines and Industrial Hygiene Division had performed under friendlier federal administrations. When explaining why Metropolitan Life, largest sponsor of the latter agency's dust research, should join the foundation, Anthony Lanza commented that

> With respect to our relationship with the Public Health Service, I can quite understand that it may not be expedient to continue further our present relationship. I do think that we should have some interest in research work in the general field of dust diseases.

The handoff went with little incident; Royd Sayers sat in on the meetings where the foundation was conceived and served on its board of trustees.

Like the federal agencies in the 1920s, the foundation did not allow its staff to testify as experts in litigation but on occasion gave defendants behind-the-scenes assistance. The new structure put full control in the hands of industry while maintaining an aura of disinterested science.[29]

Interest in industrial dust exposures ballooned as the Gauley Bridge incident belatedly drew national attention. The first publicity came in January 1935 in the *New Masses,* a Communist literary magazine. A short story in the politically influential weekly was quickly followed up with a pair of nonfiction articles. A year later, the news spread into the mainstream press, first reaching the *Pittsburgh Press* in December. In January 1936, congressional hearings were organized by Rep. Vito Marcantonio, a New Yorker who occupied a unique political niche: although elected on the Republican ballot line, he was an ardent New Dealer closely allied with the Communists. The hearings were accompanied by extensive coverage in the national press, including *Time, Newsweek, Business Week,* and the *New York Times.* Industry lost no time in reacting; public relations were added to the program of the Air Hygiene Foundation as soon as it came into formal existence on December 2, 1935.[30]

In response to the new political situation, industry reversed its position on a key legislative issue, whether silicosis should be covered by workers' compensation. In 1934, West Virginia business interests had managed to defeat a labor-backed bill providing compensation coverage for the disease, but the next year they changed course. A statute enacted in March 1935 made silicosis a compensable illness but imposed conditions so onerous that it was nearly impossible for a victim to collect. Baltimore attorney Theodore Waters surveyed the legislation of other states on behalf of the Air Hygiene Foundation's legal committee and made legislative recommendations. Waters argued, as business interests had since the 1920s, that decision making should be placed in the hands of medical experts selected for their experience in industrial hygiene. As the usual way to gain such expertise was to work for manufacturers, the effect would be to slant decision making in their direction.[31]

Another major item on the foundation's agenda was to set a standard for allowable worker exposures to silica dust. Industry wanted a limit that it could meet without severely disrupting production. Here we see a pioneering use of what was to become a common strategy: if the foundation's number could be accepted as the consensus of the technical community, employers would be protected against liability suits charging negligence. There were objections to overcome; Daniel Harrington of the Bureau of Mines contended that there was insufficient knowledge to identify a level of exposure that was truly safe. Harrington was bypassed when the standard-setting task was assigned to a Medical Committee where he, as an engineer, could not serve and his views could be ignored. This division of labor had little logic behind it; the standard was based on the ability of engineers to reduce dust levels rather than any medical determination of safety. But Harrington was in no position to make a public row; his Bureau of Mines safety inspectors had no legal authority to enter mines and depended on voluntary cooperation of mine owners to do their work.

The foundation committee, which was chaired by Anthony Lanza and counted Royd Sayers among its members, proposed a value of 5 million silica particles per cubic meter of air. The choice of this number was an exercise of discretion, carried out by a body responsible to its corporate sponsors rather than the general public. The committee recognized, in a publication intended for the foundation's membership, that it lacked "the knowledge upon which to base such thresholds."[32]

———

Meanwhile, Sayers' choice of residence turned out to have another advantage. John L. Lewis, president of the United Mine Workers, lived in a much smaller house across the street. Lewis and Sayers were well matched in their lack of scruple, and they became close friends.[33] In 1940, Sayers capitalized on the friendship and won appointment as director of the Bureau of Mines. Sayers' selection was greeted with quiet satisfaction by industry, while it allowed Interior Secretary Harold Ickes to escape a sticky political situation created by his dismissal of Sayers' predecessor. Ickes also wanted to end the Bureau's practice of communicating its findings about mine accidents in secret to mine owners.[34] Sayers at the Bureau of Mines acted as Ickes' loyal lieutenant, abandoning his previous pro-business orientation and even helping to publicize studies of black lung disease.[35] Eighteen months after the appointment, Sayers sold the Robert E. Lee house, whose historical associations fit awkwardly with the Mines director's new stance as a New Dealer. He moved to another of Alexandria's grand historic houses, the Lyceum.[36]

Sayers remained at the Bureau of Mines until a new interior secretary replaced him in 1947. While still in office as a lame duck, and (according to the syndicated newspaper columnist Drew Pearson) without checking with his superiors, Sayers sent Lewis a letter that the Mine Workers' president used to justify a national coal strike.[37] A few months later, he was rewarded by Lewis with the position of chairman of the medical board of the union's newly established health plan, which in the following years resisted recognition of black lung disease.[38]

By the time he retired from the Public Health Service in 1948, Sayers had moved on to Washington's exclusive Georgetown neighborhood. Soon

Three historic residences owned by Royd and Edna Sayers. Top left: Robert E. Lee Boyhood Home, Alexandria, Virginia. Top right: The Lyceum, Alexandria, Virginia. Bottom right: His Lordship's Kindness at Poplar Hill, Prince George's County, Maryland. (Courtesy of the Library of Congress, Historic American Buildings Survey.)

thereafter he purchased a new home, an early-18th-century plantation house on a 230-acre estate called "His Lordship's Kindness" in the Maryland suburbs. The seller was David Bruce, Ambassador to France and ex-husband of Ailsa Mellon Bruce, heiress to one of the nation's largest fortunes.[39] Sayers remained active in industrial hygiene as a consultant, and until 1963 he also worked part-time for the Baltimore Health Department with the title of Senior Medical Supervisor, Occupational Diseases.[40]

It is never easy to untangle human motivations, especially at a remove of decades. The scientists who provided rationalizations for pollution did so for many reasons. Honest mistakes and intellectual stubbornness were at work alongside conformism and careerism. But the life of Royd Sayers suggests that honest error cannot fully explain the triumph of industry's favored ideas about pollution. The recurrent theme of Sayers' scientific career is willingness to twist research to yield predetermined answers. A taste for expensive homes and a conveniently timed political conversion complete an unmistakable portrait of opportunism.

PART II | Fetching a Flood

| The Miracle Bug-Killer

As an entomologist and lover of nature, I believe that the use of aerial spraying with DDT should be reserved for serious military emergencies. DDT is such a crude and powerful weapon that I cannot help regarding the routine use of this material from the air with anything but horror and aversion.

—John Scharff, 1945[1]

THE SECOND WORLD WAR was a turning point in chemical control of weeds and animal pests. Before the war, the making of pesticides was a straightforward business. Manufacturers sold plant extracts like nicotine and simple inorganic compounds of arsenic and other toxic elements. Scientists sought to do better, and by the 1930s they had identified numerous synthetic organic compounds with properties that might make them useful for pest control. But practical applications were as yet few. War brought an urgent need for insecticides and defoliants, and intense programs of applied research were launched. By the time the conflict ended, synthetic organic pesticides were manufactured in large volumes for military use.

With the coming of peace, these synthetic pesticides quickly went into civilian use. They rapidly penetrated markets, showing that a vast business opportunity was open, and chemical manufacturers strove to develop additional categories of pest control agents. Entering the market under a regulatory regime that years of political conflict had rendered impotent to control their dangers, the synthetic pesticides were to wreak great damage on the environment.

————

Environmental protection and public health were not the concerns of the Insecticide Act of 1910, which remained in effect until 1947. The law aimed to guarantee that buyers of what were then called "economic poisons" got what they paid for. Thus it required accurate labeling and authorized seizure of adulterated or misbranded products, but did nothing to ensure anything was safe.

There was another law that could be used to protect the public against harmful pesticides. The 1906 Food and Drug Act authorized seizure of food in which added ingredients were "injurious to health." Injuries to health were not long in coming. In 1919, a Boston health inspector reported pears flecked with white spots; laboratory tests showed the spots to be arsenic compounds, residues of spraying with a pesticide, lead arsenate, that had come into wide use in moth-infested orchards. Soon fruit with high levels of arsenic was seized by local public health authorities in Boston, Los Angeles, and other cities. But federal regulators limited themselves to persuasion; the Republican agriculture secretaries of the 1920s, responsible for enforcing the food purity laws, were reluctant to take on the powerful farm lobby. The policy of persuasion had little effect, and use of arsenical pesticides continued to grow.[2]

In other countries, the government was not so lax in protecting the public. Strict standards for arsenic in food had been in place in Britain since the dawn of the twentieth century, when contaminated beer caused a severe outbreak of arsenic poisoning. A Royal Commission had been convened to search for causes and prescribe remedies. At its head was the great scientist Lord Kelvin, who half a century earlier had invented the absolute temperature scale, discovered the second law of thermodynamics, and designed the transatlantic telegraph cable. The commission's 1903 final report found that arsenic could be detected reliably with detection limits "well below" 108 parts per billion (ppb) in solids and 36 ppb in liquids. It recommended a goal of no detectable arsenic in food, with an enforceable standard of 1,080 ppb, or effectively one part per million, in solid food and 108 ppb in beer, drinking water, and other liquids. Kelvin's enforceable standard for arsenic mutated into a limit, adopted in most of Europe, on the pesticide residue on fruits and vegetables. It was referred to in the United States as the "world tolerance"— called a tolerance because a poison, ordinarily forbidden in foods, was tolerated in limited quantities due to its economic value.[3]

The U.S. government was at last forced into action by the arsenic poisoning of an English family in October 1925. Widespread inspections of American

Spraying an apple orchard, Saratoga County, New York, 1930. (Courtesy of the National Archives & Records Administration, Agriculture Historical Photo Collection.)

fruit in Britain found so much arsenic in apples that fruit sellers were fined. When British authorities threatened to ban fruit imports from America, the U.S. Agriculture Department had to do something. A voluntary system of state and federal inspection was applied to apples destined for export; fruit that exceeded the world tolerance was embargoed by the state inspectors. The creation of this system was not announced to the American public for fear of dampening domestic apple sales, but the Agriculture Department did begin to seize fruit shipments that contained too much arsenic.

The allowable level for domestic consumption was higher than the world tolerance and was at first not disclosed. Walter Campbell, about to be chosen in 1927 as the first head of the department's Food, Drug, and Insecticide Administration, explained the reason for this decision by the secretary of agriculture. The department could not openly refuse to protect public health, but enforcement of the world tolerance would ruin western apple growers.[4]

The new FDA (insecticides were removed from the name in 1930) was buffeted by criticism. Apple growers and other farmers chafed at controls, and they had strong legal weapons at their disposal—the agency bore the burden,

Inspector stopping trucks to inspect apples for lead arsenate, 1920s.
(Courtesy of the U.S. Food and Drug Administration, Historical Office.)

if there was objection, of proving in court that a seized shipment was harmful
to health. Complaints came from the opposite direction as well; concerns about
spray residues were voiced increasingly in medical journals, and in the early
'30s popular writers and newly invigorated consumer organizations assailed
the dangers of unsafe food. The double standard of allowing Americans to eat
food that was deemed unsafe for export was untenable, and the domestic tolerance was gradually reduced until it reached the world level in 1932. When
Franklin Roosevelt's left-leaning assistant secretary of agriculture, Rex Tugwell, took office in 1933, he was appalled to learn that poisons were allowed
on food, and he immediately cut the tolerance further. Loud objections from
apple growers and their senators ensued, and Secretary of Agriculture Henry
Wallace quickly ordered a partial reversal of Tugwell's reduction.[5]

With the FDA's efforts at enforcing tolerances increasingly bogged down
in court, the agency sought to put its pesticide rules on a firmer basis. The
new Food and Drug Act that Tugwell proposed in 1933 had this as one of its
objectives. The final law, enacted in 1938 after protracted struggles among a
panoply of consumer and industry interests, was a compromise—tolerances
became legally enforceable, but only after passing through public hearings.
In addition, they were subject to challenges in court. Inspectors could continue to use numbers that had not gone through this process, but when the
agency enforced these "administrative tolerances" it bore the burden of proving in court that seized foodstuffs were harmful to health.[6]

The Food and Drug Administration, meanwhile, had begun a research
program on pesticide residues. There was no way to be sure whether the

existing tolerance, derived from Lord Kelvin's study of acute poisoning of beer drinkers, was set at the proper level to prevent harm from steady consumption of smaller amounts of arsenic. By feeding low levels of lead arsenate to rats over a lifetime, the FDA sought a better basis for setting and enforcing tolerances.[7]

This study addressed an issue that bedevils environmental regulators to this day. When levels of allowable exposure to chemicals are determined, how does one take the unknown into account? Two ways of resolving this conundrum, each of which still has its advocates, were before the FDA. In ruling on smelter suits and pesticide seizures, the pro-business courts of the early twentieth century often demanded concrete evidence of harm before they would allow government to act. This legal paradigm was imported into science by Robert Kehoe in his studies of leaded gasoline. A different premise, what is now known as the precautionary principle, underlay Lord Kelvin's recommendations. His commission explained its reasoning in a preliminary report:

> ...in the absence of fuller knowledge than is at present available as to the possible effect of consumption of mere traces of arsenic, we are not prepared to allow that it would be right to declare any quantity of arsenic, however small, as admissible in beer or in any food, and we think it should be the aim of the manufacturer to exclude arsenic altogether.

The tolerance of 1 part per million was conceived as a practicable approximation to the goal of arsenic-free food.[8]

On this occasion, the clash between the two philosophies would not be resolved by the methods of science. The FDA experiments aroused the ire of Rep. Clarence Cannon of Missouri, an apple grower who thought residue regulation was "nonsense." Cannon became chairman of the subcommittee that funded the FDA, and Congress soon acted. The 1937 appropriations act forbade the agency to do research on the health effects of spray insecticides and deleted $50,000—a substantial sum in a year when the government's entire cancer research budget was $115,000—from its budget. The experiments were halted with the slaughter of 5,000 rats at the end of the fiscal year on June 30.

Cannon was not finished. If lead and arsenic were to be openly allowed in food, a scientific justification had to be concocted. For this he turned to Royd Sayers. The matchmaker who brought together the legislator and the Public Health Service was Felix Wormser, secretary of the Lead Industries Association and orchestrator of the defense of lead paint. In his efforts to protect the lead makers' markets, Wormser was not fastidious. He used his control of industry research funding to direct scientists away from sensitive topics while striving to keep warnings about the dangers of lead paint on toys and cribs out of the popular press. Wormser's PR campaign was too much even for Robert Kehoe, who without directly confronting the industry warned repeatedly of the dangers lead posed to infants.

Before taking any action, Cannon first arranged with Sayers a study plan that met his approval. The congressman then appeared as a witness before the appropriations subcommittee with jurisdiction over the Public Health Service. That organization was instructed to investigate the topic that the FDA had been told not to look at, using the $50,000 that had just been freed up. The study was carried out under the direction of Sayers' subordinate Paul Neal. Neal was chief of the Industrial Hygiene Division's laboratory, the unit that was transferred along with Sayers out of the Bureau of Mines at the end of the Hoover administration.[9]

Sayers' research protocol reprised the techniques used to slant earlier studies. As in the leaded gasoline investigation, broad conclusions about public safety were drawn from narrowly limited observations. A question about chronic diseases and lifelong exposures—in this case, how much lead and arsenic the fruit-eating public can safely consume over a lifetime—was answered by examining exposed workers for symptoms of acute poisoning. Former employees were excluded—a procedure that, due to the nature of orchard work, biased the sample even more than it had in Sayers' coal mine study. The workers with the highest pesticide exposure were apple pickers, who work on ladders. But acute lead poisoning causes the "shakes," a condition that leads its victims to seek employment on solid ground. By this indirect means, workers who displayed the symptoms of poisoning were culled out from the study population composed of active apple pickers.[10]

This was not all that was done, it seems, to ensure that the study reached Rep. Cannon's desired conclusion. Volunteers were recruited in the apple-growing region around Wenatchee, Washington, through radio and newspaper announcements and by talks at "grange meetings, service clubs, and chambers of commerce." The horticultural inspector—a locally based state official whose main responsibility is eradication of insect pests—was "most helpful in sending subjects in for investigation." FDA scientists, once they learned of this, were convinced that the apple growers had screened the study participants to weed out the sick.[11]

Adverse findings still could not be avoided and had to be downplayed. Buried 57 pages into the report text is the information that "Diagnoses of low-grade lead arsenate intoxication were made" on seven workers; the summary at the front of the document distorts this into

Some physicians may interpret these cases as minimal lead arsenate intoxication. However, as regards lead, these cases do not come up to the criteria...[12]

The Industrial Hygiene Division's report was sent to the FDA in June 1940. Royd Sayers had just left for the Bureau of Mines, but his replacement by James Townsend did not bring any change in the division's behavior. The pesticide report recommended more than a doubling of the arsenic tolerance and a tripling of the lead tolerance. Walter Campbell, still head of the FDA, had long subscribed to Lord Kelvin's goal of allowing no residues at all in food. He was so shocked when he got the report that he at first had the document literally locked up. His opinion of the report was shared by the FDA staff, who were surely not reassured when Neal defended his conclusions with the argument that combining arsenic with lead might make the lead less poisonous. The agency commissioned an independent review by Anton Carlson, who had chaired a National Academy of Sciences committee overseeing the earlier rat experiments. Carlson stated flatly that the work "furnishes no scientific basis for permitting more lead and arsenic on apples and pears." The final decision was in the hands of Administrator Paul McNutt of the new Federal Security Agency, which included both FDA and PHS. McNutt sided with the Industrial Hygiene Division and the growers, and in August 1940 the administrative tolerances were increased.[13]

Campbell then sought to call hearings so that the tolerances, if too high, would at least be enforceable. The Industrial Hygiene Division delayed another year until its report was deemed ready for publication. At that point the growers and shippers, happy with the situation as it was, took up delaying tactics. Delay in wartime was easy to obtain, and enforceable tolerances for arsenic and lead were not set until 1950. By that time, it made little difference; arsenical pesticides had been eclipsed in importance by new synthetic compounds.[14]

———

Dichloro-diphenyl-trichloroethane, or DDT, came to the attention of the U.S. army because, fighting a war in Africa and the South Pacific, there was a need to kill malaria-carrying mosquitoes. The nation's scientific capabilities had been mobilized for the war effort, and the task of controlling insect-borne diseases was assigned to the Agriculture Department's entomology laboratory in Orlando, Florida. The Orlando scientists recognized that war conditions required different approaches than peacetime: solutions should be quick, rather than optimal, and they should be usable worldwide rather than tailored to local conditions.[15]

The first problem they tackled was not malaria, but typhus. This disease was carried by lice in soldiers' uniforms and bedding; it had traditionally been controlled by steaming to kill the lice. This method was slow and hard to use on the battlefield; the army hoped to replace it with a louse-killing powder. The researchers screened some eight thousand chemicals; after a series of progressively more rigorous tests they settled on pyrethrum, a material extracted from dried chrysanthemum flowers. Accelerated toxicity tests with animals indicated that the powder would cause no immediate harm to users' health. Such testing was understood to fall short of what would be needed in peacetime, but 20,000 soldiers had already been lost to typhus in Africa. Herbert Calvery of the Food and Drug Administration weighed the risks and concluded that "the benefit...will far outweigh any harmful effects resulting from their possible toxicity." Pyrethrum powder became the military's standard louse killer after the laboratory recommended it in August 1942.[16]

Orlando's next assignment concerned mosquitoes that carried malaria, a disease that in the Pacific Theater caused more casualties than the enemy. The traditional methods of malaria control by draining swamps or covering standing water with oil were workable only behind the front; the army sought a way to kill adult mosquitoes on territory that had not yet been invaded. The obvious choice was to spray an insecticide from the air. The technique of mixing insecticides with propellant gases and dispersing them as aerosols had been developed the previous year; it was already in use in the form of one-pound "aerosol bombs," the predecessor of today's spray cans, used to kill mosquitoes indoors.[17]

Unfortunately, the insecticide used against mosquitoes was the same pyrethrum used to fight typhus; it was imported and supply was short in wartime. The Orlando scientists once again started screening chemicals. A sample of DDT arrived in November from the Swiss company Geigy. Paul Müller, a Geigy chemist, had found DDT to be effective in killing insects, and safety testing had shown it to be relatively nontoxic to man and animals. Geigy was preparing to manufacture DDT in wartime Germany as well as in Switzerland, and it licensed the compound to German chemical manufacturers as well as to the Americans and British. But it was the Allies, forced unlike the Germans to fight in tropical conditions, who gave the pesticide its fame.

The American tests quickly proved DDT to be ideal for the army's purposes—it was effective and long-lasting, and it could be manufactured in the United States. Safety testing began immediately. Results soon arrived from the Food and Drug Administration, and they raised troubling questions. When fed to animals, the compound was found to cause nervousness,

A soldier demonstrates how to apply DDT to kill the lice that were carriers of typhus. (Courtesy of the Centers for Disease Control, Public Health Image Library.)

convulsions, and, with large enough doses, death. In one positive finding, it was not absorbed through the skin in powder form, but when dissolved in solvents or ointments it passed readily through rabbit skin and caused severe illness. A war was on, and the need was urgent; DDT was approved for use as a louse powder in May 1943.[18]

But DDT was not yet approved as a spray. The crucial experiments that opened the door to its widespread use were directed by Paul Neal, the Industrial Hygiene Division scientist who had suggested that lead becomes less toxic when it is combined with arsenic. Neal's forte was not scientific consistency. Three years earlier, he had extrapolated from observed effects of inhaling lead and arsenic to calculate how much of those substances could be safely eaten. Now he ignored demonstrated ingestion hazards when judging the safety of DDT inhalation.

As with lead arsenate, the broad conclusions that Neal drew from his DDT inhalation tests could be supported only by a selective reading of his data. The experiments showed that DDT aerosols often killed mice, but did little harm to dogs or monkeys. Two human beings exposed to DDT in the same form showed no evidence of poisoning. Neal's report, issued in September 1943, declared that DDT could be used safely as aerosol, dust, or mist; spraying was approved that same month.[19]

Production ramped up quickly; as it turned out, the supplies arrived just in time. A winter typhus epidemic in Naples was quelled with the help of DDT, winning the chemical wide acclaim. By the summer of 1944, the infection rate among soldiers in the Southwest Pacific had fallen 95%. DDT was hailed in hundreds of news articles as one of the great scientific discoveries to emerge from the war. With the press eager for a simple, compelling story, coverage often ignored the hazards. The *Chicago Tribune*'s encomium to a substance "harmless to humans and warmblooded animals" was typical.[20]

Among scientists, too, there was excitement about the potential future benefits of DDT, but doubts remained about safety. At a May 1944 press conference, government entomologist F. C. Bishop gave an implicit warning. The persistence of DDT and its ability to kill many different insects, qualities that had made the chemical so useful to the military, posed dangers in civilian use. Residues could be left on food, and new pest problems could be created by killing insect predators. Food and Drug Administration researchers warned of the compound's toxicity in the diet and its tendency to accumulate in body fat and milk. Researchers at the National Institute of Health reported in July 1944 on effects on nerve cells, brains, muscles, kidneys, and livers. Neal of the Industrial Hygiene Division, meanwhile, trumpeted the safety of DDT. On November 2, one day after the War Production Board warned that more research was needed before DDT could be used commercially, he gave a speech declaring DDT to be safe for humans.[21]

Concern about whether DDT might cause cancer arose in several places and quickly gained attention. The Army surgeon general wrote to Surgeon General Thomas Parran in August 1944 to ask for an investigation by the National Cancer Institute. Parran agreed, but warned that testing would take several years. Lack of staff caused delay; at war's end the work had not yet begun.[22]

Experience with spraying in the Pacific, where widespread destruction of fish and shellfish was reported, magnified the concerns of entomologists and fish and wildlife biologists. Tests by the Fish and Wildlife Service in Maryland during the summer of 1945 showed that small doses of DDT killed fish, and larger doses affected birds as well. John Scharff, a prominent British malaria expert who supported the use of DDT in wartime, warned that the substance should be reserved for military emergencies. He reacted to the prospect of widespread aerial spraying with "horror and aversion." These concerns were not hidden from the public; nature writer Edwin Teale warned of a "conservation headache of historic magnitude," and an April 16 article in *Time* reported that DDT, although full of promise, "is not yet safe for general use."[23]

On April 20, 1945, the Army and the Public Health Service issued a ban on aerial spraying of DDT for civilian purposes. Small-scale use on individuals or indoor spaces was permitted, but outdoor spraying required a special exception that was granted for only seven applications, all on military bases, in the next three and a half months. A press release explained the rationale for this decision:

> Much still must be learned about the effect of DDT on the balance of nature important to agriculture and wild life before general outdoor application of DDT can be safely employed in this country. It may be necessary to ignore these restrictions in war areas where the health of our fighting men is at stake, but in the United States such considerations cannot be neglected.[24]

But the authority by which this edict was issued derived from the war and the wartime system by which materials in short supply were allocated to satisfy war needs first. The rapidly increasing production of DDT would soon exceed the requirements of the army and navy. By the summer of 1945, a small surplus was foreseen in the fourth quarter. The DDT Producers Advisory Committee, a group of chemical company executives whose advice would have strong influence on the War Production Board, met with staff from that agency and the Agriculture Department's Bureau of Entomology on July 26.

The Bureau of Entomology argued for continued controls: use should be allowed only when government scientists had determined the spraying was necessary, would not leave harmful residues, and would not cause ecological upset. The manufacturers were divided. One asked the War Production Board to maintain rigid supervision; uncontrolled use would be dangerous because "we have not yet established the safety controls." Another view prevailed, however. As soon as there was a "considerable surplus," companies would be free to sell to anyone in any quantity.[25]

Neither the reasons for the decision nor the identities of the pesticide manufacturers holding different views are recorded in the surviving documents, but the outcome of the meeting reflected the pesticide laws of the time. Labels had to be accurate, and foods bearing excessive pesticide residues could be seized, but no government agency had the power to stop the sale of an accurately labeled pesticide. The labeling rules were administered by the Insecticide Division, which had stayed behind in the Agriculture Department when the remainder of the FDA was transferred in 1939 to the Federal Security Agency.

In April, the month of the ban on civilian DDT spraying, instructions for labeling DDT for civilian use had been issued. A listing of ingredients was not required unless inactive ingredients were added. There was little the FDA, whose scientific knowledge was far too scant to hold up in public

The first use of aerial spraying with DDT to control forest insects in the western United States, Latah County, Idaho, 1947. (Photo by P.C. Johnson, USDA Coeur d'Alene, Idaho, Forest Insect Laboratory, 1947. Courtesy of the archives of the Western Forest Insect Work Conference.)

hearings, could do to control the new insecticide's use. The agency decided in November to take the one number that had made it through the hearings process, a value of 7 ppm for fluoride, and use it as an administrative tolerance for DDT.[26]

The producers, Sievert Rohwer of the Bureau of Entomology told the July 26 meeting, "will have a great responsibility as well as an opportunity. He expressed the hope that they would use their opportunity wisely. He stressed that there was a great deal that was still unknown on how to formulate DDT insecticides so they could be used safely, effectively, and to the interest of the user and the public good."[27]

Within weeks, DDT was on the market and used with striking effect. A storm of publicity followed, enthusiastic case reports mixing with sober warnings of the dangers of overuse. In the absence of regulation, the excitement quickly overshadowed the cautions. Edmund Russell, the historian of the development and approval of DDT, describes the aftermath:

> At first, market forces limited sales by leading risk-averse manufactur-
> ers, which tended to be the larger companies, to factor in the cost of

potential lawsuits and hold off on sales. But smaller companies, apparently less daunted by the risk, went to market. As it turned out, DDT did not produce much visible, short-term damage, and the larger companies followed the small ones into the market. Sales soared.

Scientific criticism was soon overwhelmed by commercial success. Paul Müller, the Swiss scientist who discovered DDT's utility as an insecticide, was awarded the Nobel Prize for chemistry in 1948.[28]

Under wartime conditions, Geigy had been compelled to adopt a policy of licensing its patents on DDT to all comers on favorable terms. Prices fell as new producers entered this uncontrolled market, and large, risk-averse companies found it hard to compete. DuPont never recovered its investment in DDT; by 1954 it was losing three cents on each pound it sold and ceased production of the chemical. Geigy's royalties gave it little economic advantage, and in the end the company was disappointed with the product's financial yield.[29]

As the Second World War drew to a close, government pesticide regulators saw that a new insecticide law was needed. The Food and Drug Act of 1938 had rewritten the law on residues, but otherwise the 1910 Insecticide Act remained the governing statute. With a flood of synthetic compounds ready to hit the market at war's end, chemical manufacturers were about to set in motion a force that could not be controlled by the competitive marketplace.

Edward Griffin of the Insecticide Division began drafting a new law in 1944. From the beginning, his premise was that no bill could pass without the full support of the pesticide industry. This could not be faulted as a political calculation; the arsenic residue debacle of 1937–38 was the work of the Congress elected in Roosevelt's 1936 landslide, and business influence over the legislative branch had only increased since.

In the new legislation, the Agriculture Department offered the industry a formula to harness the genie that was about to be released. It sought authority to register and approve new pesticides before they could be sold. Controls on advertising were requested as well. There was also a requirement for coloring of powdered pesticides, aimed at overcoming the problem of accidental poisonings by chemicals mistaken for flour or sugar.

The manufacturers declined the opportunity to bring pesticides under comprehensive regulation. True, they had long supported the modest regulatory system enacted in 1910, which allowed the sale of any product, no matter how dangerous, as long as it was honestly labeled. During the war years, they had even lobbied Congress for increased appropriations so that the division could enforce the rules more effectively. But the labeling law served to shield legitimate producers against fraudulent competition as much as it protected the public. The industry still resisted regulation of its products' effects. They were willing to live with the coloring rule, but would go no further.[30]

The effective oversight that would have been needed to bring the new synthetics under control was thus precluded; Griffin nevertheless hoped for incremental progress. As soon as he had agreement on a draft bill within Agriculture, he presented it to the industry and began making changes to meet criticisms of the registration system. In November 1945, a bill was submitted that went partway to meet the objections. A pesticide that the Agriculture Department refused to approve could still be sold, "registered under protest," with a warning issued to the manufacturer.[31]

Congressional hearings in February 1946 revealed that consensus had not been attained. Although manufacturers agreed to registration in principle, they remained skeptical; on the other side, some members of Congress questioned the concept of registration under protest. Conservation groups, already concerned about ecological effects of the synthetics, did not speak, but their concerns were reflected in testimony by state regulators. After the hearings, the bill was redrafted to meet industry objections and to bring herbicides within its scope, but it failed to pass before adjournment.

The new Republican Congress of 1947 quickly took up the issue. Industry objections to federal registration had disappeared after states, at the prompting of the Agriculture Department, began to pass their own registration laws. Manufacturers preferred a uniform system of weak controls to a patchwork of state rules. But new objections arose to limits on advertising; these were dropped. The amended bill sailed quickly through and the new law was signed by President Truman in June.[32]

Even the title of the new statute displayed its friendliness to industry. Politicians hoping to trumpet their legislative achievements would never have devised a moniker as ungainly as the Federal Insecticide, Fungicide, and Rodenticide Act. Such a name could originate only from the chemical manufacturers. They had long labored to replace words like bug-killer and rat poison with the scientific-sounding terminology of this-icides and that-icides.

The ineluctable fate of euphemism lay in the future of these words too. The companies that rebranded economic poisons as pesticides now prefer to call their products "crop protectants."

| CHAPTER 6 | Wilhelm Hueper and Environmental Cancer |

Confidentially, I believe that there may be something to justify these statements. While the incidence of pulmonary carcinoma among our chrome workers has been very low, it is still significantly higher, in fact many times the incidence in the general public.

—Harry Heller, Mutual Chemical Company, 1938, about an assertion in the *Journal of the American Medical Association* that chrome dust can cause lung cancer.[1]

THE STORY OF ENVIRONMENTAL cancer in the mid-twentieth century is inextricably linked with the name of Wilhelm Hueper (1894–1979). Hueper was a scientific pioneer in a field that did not fully flower until his career was at an end. Beyond that, his story epitomizes the era's struggles over environmental protection. Hueper's stubborn insistence on the dangers of modern industry made him a bête noire for chemical manufacturers even before the appearance in 1962 of Rachel Carson's *Silent Spring*, for which he was a principal scientific source. In the history of the industry's often successful efforts to suppress Hueper's research, one sees the sharply etched image of behavior that more often can only be glimpsed through a fog of evasions and excuses.

The 29-year-old Hueper arrived in America in 1923 with what would have seemed an unexceptional German middle-class background. His father, a railroad official in the small city of Schwerin in Mecklenburg, had served on the municipal council; Hueper describes him as "a liberal having a well developed social conscience." A year before the First World War, Hueper entered the university to study medicine. At the outbreak of war, his patriotism overcame pacifistic inclinations and he immediately volunteered for service. Four years in the army, ending with four months as a prisoner of war, greatly reinforced his aversion to warfare, but Hueper's political outlook remained essentially centrist. In the spring of the revolutionary year of 1919, he joined with other students in militias that defended the infant German republic against far-left uprisings; by summer, as the militias aligned with the far right to seek a restoration of the empire, he chose to stay home.[2]

Hueper finished medical school in 1920 intending to become a gynecologist. After a year's study of pathology in Berlin, Hueper found himself in 1923 ready to marry but without steady employment at the peak of Germany's hyperinflation crisis. He accepted a job offer from a Chicago hospital, but the job proved on his arrival to be less than advertised. He managed instead to find a position as a pathologist at Loyola University's teaching hospital and brought his bride over. He soon resumed the research he had begun in Berlin; among numerous publications was a paper that prefigured his later work, arguing that automobile pollution and industrial chemicals were responsible for a sharp increase in the incidence of lung cancer. He also wrote abstracts of German-language scientific papers for the American Medical Association's pathology journal.

Denied a hoped-for promotion, Hueper moved to Philadelphia to work at the cancer laboratory of the University of Pennsylvania. The laboratory was supported by Irénée du Pont, vice chairman of the chemical company, and directed by Irénée du Pont's personal physician, Ellice MacDonald. Visiting the DuPont dye works with MacDonald, Hueper was told by chemists of the plant's use of beta-naphthylamine and benzidine. These compounds had been used in the large-scale production of synthetic dyes in Germany since the 1880s, and Hueper knew from his abstracting work that German researchers suspected them of causing bladder cancer. He sent a memo about this danger to Irénée du Pont.

The memo was not the first warning to the DuPont company—a chemist in the company's research department had learned of the problem from a British colleague in 1928. DuPont at the time had seen no cases of bladder cancer in dye workers; cancer emerges after a delay of years, and the company

Dr. Wilhelm Hueper, pioneer of environmental cancer research, at his microscope in the pathology laboratory at the Cancer Research Laboratories of the University of Pennsylvania, 1931. (Courtesy of the National Museum of American History.)

had begun to produce dyes only in 1917. There was not long to wait; by the fall of 1932, 20 bladder tumors had appeared. DuPont's medical director, George Gehrmann, was sent on a tour of Europe the next autumn to seek information and help. On his return Gehrmann developed a plan for improved hygiene in the dye works and recommended that the company establish a toxicology laboratory.[3]

Gehrmann's proposal was quickly approved. The need for testing was increasing rapidly as new chemicals entered the market; and the onset of the New Deal made it uncertain whether the confidential arrangements whereby such studies had been done under contract by the Bureau of Mines could continue. Dow and Union Carbide also established in-house toxicology laboratories around this time.[4]

In the meantime, Hueper's outstanding character trait, his obstinate outspokenness, had emerged. After frankly expressing his dissatisfaction with MacDonald's scientific standards, he soon learned that the director planned to dismiss him. Facing unemployment at the nadir of the Great Depression, he undertook what was surely the least creditable action of his life, a trip to Germany to seek one of the medical positions that had been opened up by Hitler's dismissal of Jews and political opponents from the profession. His initial job inquiries did not bear fruit, however, and what he learned of the new regime dissuaded him from further pursuit of this objective. He returned to Philadelphia at the end of 1933, knowing that his future lay in the United States.[5]

After passing some months in ill-paid and itinerant pathology positions, Hueper received a letter asking him to join DuPont's new toxicology laboratory to pursue the problem of which he had issued an early warning, bladder cancer among dye workers. This opportunity to engage in fundamental research was Hueper's dream; on opening the letter, he ran home and told his wife they were moving again. His experience at DuPont was to determine the course of his subsequent career.

Hueper began work in November 1934 as chief pathologist of what was named the Haskell Laboratory. In experiments on dogs, he was able to show that beta-naphthylamine, and not other chemicals used in the plant, produced bladder tumors similar to those observed in the diseased employees. This finding, published in 1938 in the *Journal of Industrial Hygiene and Toxicology*, was a landmark in cancer research, the first experimental demonstration that a synthetic organic chemical causes cancer.

By the time the article appeared in print, Hueper's employment at DuPont had ended on bad terms. On a plant visit early in his dye research, he learned that the area of his visit had been specially cleaned for the inspection and insisted on seeing the much dirtier remainder of the operation. His immediate written complaint went to Irénée du Pont, and Hueper was never allowed to visit the dye works again. Later, when the results of the experiments became clear, DuPont issued a press release announcing the discovery and ascribing it to the laboratory directors; Hueper visited the local newspaper to assert his own claim to credit and kept the release out of the paper. Soon afterward he was dismissed.[6] DuPont quietly financed continued research on dye chemicals by Robert Kehoe, not publishing the findings even when they pointed to cancer hazards from additional chemicals.[7]

———

As Hueper was busy with his dog experiments, other widely used materials began to emerge as carcinogens. One was the element chromium. This metal, which in nature is almost always found in its trivalent chemical form, is extracted by converting it into the hexavalent form, which dissolves easily in water. The main commercial products are purified chromate or bichromate salts, usually sold in powder form. Because the element is extracted from the ore by oxidation—the opposite, from a chemist's point of view, of the reduction process in a copper or iron smelter—and the final product is a salt, chromium production was part of the chemical industry.

Chromium plants had long been known as unhealthy workplaces—employees characteristically developed holes in the bone separating the two nostrils—but in the 1930s a new danger emerged. A report describing fatal lung cancers of two workers at a chromium plant appeared in a German medical journal in 1932. A second report in 1935 raised the number of lung cancer victims at this factory to seven.

The German chemical giant I.G. Farben sent the article to the Mutual Chemical Company, the largest U.S. chromium producer. The article was read carefully by its recipient, Herbert Kaufmann, who was manager of the company's Jersey City plant and the son of its principal owner and president. But it elicited little immediate action other than a request that I.G. Farben "have correspondence concerning this general topic addressed directly to the writer."[8]

Three years later, there appeared a publication on the subject that was harder for Mutual to ignore. It was in the pages of the *Journal of the American Medical Association*, still edited by the same Morris Fishbein who had led the Chicago investigation of refrigerator poisonings eight years earlier. A doctor in Jersey City wrote in to ask whether his patient's lung cancer could be caused by his employment in a chromium factory. The editors replied in the affirmative and provided five journal references. Mutual's president, H. M. Kaufmann, sent an inquiry to both of the company's plants. Harry Heller, comanager of the Baltimore works, replied that he was familiar with three of the articles listed by the journal, and added "confidentially" that lung cancer

Mutual Chemical's Baltimore plant in the 1940s. (Courtesy of Douglas Janney.)

among company employees was much more frequent than in the general population. Kaufmann wrote back that the company should make no statement on the matter, adding that the plant managers should take steps to improve working conditions.[9]

Another carcinogen identified in this period was asbestos. This discovery emerged from industry-sponsored research supervised by Anthony Lanza, the former chief of the Public Health Service's Industrial Hygiene Division who had gone to work for Metropolitan Life Insurance. After a 1930 study sponsored by Metropolitan Life found high rates of lung disease among asbestos miners in Quebec, Lanza denied permission for publication and arranged for the asbestos industry to join Metropolitan in sponsoring research at the Saranac Laboratory, a lung research institution in upstate New York. Within a few years, these experiments showed that cancer was a problem as well, a conclusion that had also begun to appear in German medical literature. By 1943, an interim report of Saranac's work noted that 81.8% of animals exposed to asbestos developed lung tumors.

In 1948, the new director of Saranac, Arthur Vorwald, completed the asbestos report and submitted it, with a discussion of cancer included, for review by the sponsors. The sponsors committee met in early November and decided that the cancer section should be deleted. Lanza, who was on the board of Saranac, communicated this decision to Vorwald the next month. The draft report, which had been distributed in numbered copies to the sponsoring organizations, was recalled, and the revised report without a cancer section was published in 1951. When Vorwald in 1953 sought to further explore the relationship between asbestos and cancer, Lanza reportedly forced him to resign from Saranac and flew to California to block his appointment to a university position.[10]

———

Meanwhile, Hueper had found a job with a drug manufacturer, William R. Warner, that left him time for his own projects. In the summer of 1938, he began work on what turned into a 896-page tome entitled *Occupational Tumors and Allied Diseases*. Published just after the New Year of 1942, this work collated and critically evaluated a scientific literature that was extensive but widely scattered and diverse in its methods. The response to Hueper's claim that occupational cancer had "an ever-increasing importance as a part of the problem of cancer in general" was mixed. While Hueper's accomplishment in organizing a vast corpus of knowledge won praise, reviewers in several medical journals felt that Hueper exaggerated the prevalence of industrial cancer. They faulted in particular his willingness to draw general conclusions from medical case reports.[11]

Wilhelm Hueper during his years at the National Cancer Institute, where he struggled against industry obstruction of his work. (Courtesy of the Office of National Institutes of Health History.)

Notwithstanding controversy over his medical opinions, Hueper's scientific credentials were unquestioned. He was invited to write unsigned editorials for the *Journal of the American Medical Association*. Morris Fishbein was a political conservative, a fierce opponent of national health insurance, and an unrelenting critic of whatever he thought unscientific, issuing broadsides against homeopathy and chiropractic as well as out-and-out quackery. But Fishbein, leader of the 1930 inquest into refrigerants and coauthor of an industrial medicine text, had no qualms about Hueper's research.[12]

Hueper's next great opportunity came from the government. The National Cancer Institute, established in name but barely under way before the war, was staffing up. Surgeon General Thomas Parran, a committed New Dealer appointed by President Roosevelt in 1936, had systematically expanded the scope of the Public Health Service's activities, and public health agencies could point to impressive achievements in the control of such diseases as tuberculosis and syphilis. It was in this spirit that the NCI was initially conceived. It was to have two coequal branches, one for cancer research and the other for cancer control. The head position in a subdivision of the Cancer Control Branch, the newly created Cancerigenic Studies Section, was offered to Hueper at the end of 1947.[13]

Aided by only a minuscule staff, Hueper unleashed a storm of activity. Initially unable to undertake experiments while awaiting construction of the section's animal laboratory, he devoted his first year to epidemiological research and publicizing the dangers of environmental cancer.[14] The Public Health Service soon published two important documents: a 69-page survey of "Environmental and Occupational Cancer" intended for scientific readers and an 18-page booklet on "Environmental Cancer" for the general public. Hueper insisted that the problem posed by cancer-causing chemicals extended far beyond the factory gate into the general environment. Identifying the most dangerous emerging cancer hazards as radioactive substances, petroleum refining, and synthetic organic chemicals, he warned in his popular pamphlet that

> This continued exposure to the environmental cancer danger of workers, technicians, scientists, management personnel, and even the populations of communities near the plants, constitutes an urgent public problem....Safety procedures...should include not only protection of the individual workers by enclosing the manufacturing processes, but also protection of the whole plant and of the community at large by preventing the escape of carcinogenic wastes into the atmosphere, the water, or the soil.[15]

Seizing on the cancer control mandate, Hueper took on a wide spectrum of industries and chemicals during his first years at NCI. In his later writings, one finds mention of studies, plans for studies, and plant visits involving bladder cancer among dye workers, scrotal and lung cancer among paraffin pressers at an oil company, lung cancer among railroad workers, lung cancer from asbestos, skin cancer at a tar paper factory and a tar refinery, skin and lung cancer at copper smelters, and nasal and larynx cancers at isopropyl alcohol plants. In addition, a survey of cancer patients' occupations was launched, aimed at unearthing as yet unknown hazards.[16]

But the political winds shifted at the Public Health Service soon after Hueper's arrival. The service was under the umbrella of the Federal Security Agency, predecessor to today's cabinet-level Department of Health and Human Services. In 1947, President Truman appointed as the new head of the agency Oscar Ewing, a lawyer and Washington dealmaker who had served as general counsel of Alcoa Aluminum and vice chairman of the Democratic National Committee.

Ewing quickly clashed with Surgeon General Parran. In an oral history interview given long after his retirement, Ewing recounted that Parran had resisted his request that a research grant be given to the doctor who had successfully treated his wife's high blood pressure. Parran would not interfere with the committees of scientists that awarded grants at the National Institutes of Health; Ewing responded by holding up approval of all appointments to the committees. As vacancies piled up, quorums could not be assembled and some of the institutes were unable to function. An angry Parran stormed into the Administrator's office and said "You've got to approve these appointments."

"Well," Ewing countered, "I'm not going to until you can do something about this grant for Dr. Kempner."[17]

Dr. Kempner's grant was approved a few days later, but the incident was not forgotten. Parran's term was to expire the following May, and Ewing announced in February that he would not be reappointed. The surprise firing made waves. Parran's support within the Public Health Service was strong, while his advocacy for national health insurance had made him a target of criticism from Morris Fishbein at the American Medical Association and from the Republicans who controlled Congress. Parran was replaced by Leonard Scheele, who had risen to be director of NCI at the age of 39 on the strength of his administrative skills and successes in obtaining research money from Congress. Scheele, Ewing commented, was "just as cooperative as anyone could be."[18]

In the new climate, Hueper's work quickly encountered opposition. Within the Public Health Service, he came under fire from the Industrial Hygiene Division. When his scientific survey sold out a printing of 10,000 copies, permission to reprint was denied after he refused to make changes that organization demanded. He clashed as well with corporate medical directors such as George Gehrmann of DuPont and Robert Eckardt of Esso. Accused of being a Nazi sympathizer, he won clearance from a Loyalty Board; he was next said to exhibit communistic tendencies.[19]

The uranium mining industry, newly created by the atomic bomb project of the Second World War, engaged Hueper's immediate attention. Knowing of high rates of lung cancer among European uranium miners, he proposed in April 1948—when the manifestation of the disease was still in the future, as DuPont's bladder cancers had been when he warned of them years earlier— that research should be initiated by the Colorado Health Department. The Atomic Energy Commission's chief medical officer, Shields Warren, voted against this project on the grant-making committee; defeated, Warren wrote to the Surgeon General seeking to have the grant canceled. The project went forward, but without Hueper's participation.

Hueper's next run-in with the atomic energy establishment was in 1952, when he was invited to speak on the subject to the Colorado Medical Society. Ordered by his superiors to delete reference to lung cancer in German and Czech uranium mines, information that had been known to scientists since 1879, he refused with a flourish. He withdrew as a speaker by mailing the text of the paper to the president of the society, with a cover letter explaining that he would be unable to appear because he had been censored.[20]

————

The target that proved to be Hueper's undoing was chromium. His attention was drawn naturally to this industry, where a cancer hazard had been identified before the war. Hueper's work on chromium began before he joined NCI, in 1946, when he was asked to consult at the Baltimore plant of the Mutual Chemical Company. At this point he crossed paths with two men who epitomize the opposite poles of the chemical industry's response to environmental hazards: Anthony Lanza of Metropolitan Life and Omar Tarr, Mutual's technical director.

If Hueper's encounter with Mutual Chemical were a drama, Omar Tarr (1892–1955) would make by far the most interesting character. The playgoer might see Hueper and Lanza more as archetypes than as human beings; Tarr was a man who struggled in trying circumstances to balance conflicting obligations to his company, to humanity, and to the truth. The first in his family to attend college, he received a chemical engineering degree in 1916 from the University of Maine. After joining Mutual in 1920, he rose quickly to be comanager of its Baltimore plant and later became a company vice president. His invention of a more efficient method of extracting chromium from ore, using rotary kilns in place of hearth furnaces, improved efficiency so much that Mutual stayed prosperous through the depression. A short, spare man whom subordinates addressed as "Mr. Tarr," and a church elder, he had a manner that fit the stereotype of the Maine Yankee. Indeed, while raising his family in Baltimore, he took care not to lose touch with his roots. He vacationed in Maine and sent all four children back to the state university in Orono.[21]

War production, with long hours, equipment shortages, and the hiring of inexperienced personnel, had exacerbated the chromium dust problems at Mutual. A decision by the U.S. Employment Service to warn prospective employees of the bad working conditions at Mutual triggered an investigation by the Industrial Hygiene Division of the Public Health Service. Initial results were reported to the company at the end of 1945; with another death from lung cancer recorded during the year at Mutual's Jersey City plant, the severity of the health problems was only confirmed.[22]

Hueper's consultation work was the first step Mutual took in 1946 to address the cancer problem. He advised the company to study the German experience and undertake an epidemiological study.[23] Omar Tarr undertook a visit to the German chromium plants, where he consulted with their managers and medical personnel and was advised that the association between

The executives of Mutual Chemical Company, late 1940s. From left: George Benington, president; Jerome Dohan, Jersey City plant manager; Harry Heller, Baltimore plant manager; Omar Tarr, technical director. (Courtesy of Douglas Janney.)

chromate and cancer was considered proven fact. On Tarr's return, Leroy Gardner of the Saranac Institute was hired as a consultant; but Gardner's death in October cut his work short and Anthony Lanza was brought in as a replacement.[24]

Meanwhile, a compensation claim had been filed on behalf of the employee who had died of lung cancer in 1945.[25] Theodore Waters, the attorney who had helped lead the Industrial Hygiene Foundation's legal team in its efforts to repel compensation claims for silicosis, worked with Lanza to organize Mutual's defense.[26] On February 1, 1947, Lanza and Waters met with Mutual executives to discuss a problem whose nature their written communications— obviously prepared with the possibility in mind that they would be exposed in litigation—took care not to identify.[27]

On Lanza's recommendation, an epidemiological study of the company's employees was undertaken by Willard Machle, an industrial hygiene consultant who had worked for a decade at Robert Kehoe's Kettering Laboratory.[28] Six decades later, the precise motives for commissioning Machle's investigation are difficult to pin down. Was it to some degree a genuine search for knowledge? Or was Mutual simply trying to postdate the discovery of the cancer hazard[29] in order to bolster its defense against claims that it had been negligent in exposing workers to chromium dust?

Machle found, as the company already knew, that the workers had a much higher lung cancer rate than the general population. The study was then expanded with the assistance of another doctor, Frederick Gregorius, to include the country's other five chromium manufacturing plants. Machle and Gregorius published their results in August 1948. It was revealed that workers in these factories had a death rate from lung cancer 25 times higher than the general public.[30] The Industrial Hygiene Division reacted to this alarming news in the all too common manner: it proposed another study. Planning for a comprehensive investigation of the chromium industry stretched on into the next year. Lanza, having left Metropolitan Life by now, was retained directly by the industry to be its liaison with the new government project.[31]

Notwithstanding the governmental lassitude that Mutual was actively abetting, the company took action to reduce chromium exposures. Under Tarr's direction, Mutual embarked in 1948 on the design of a new plant that would replace both of the existing factories. The new facility was intended not only to remedy the problem of occupational chromium exposures, but also to reduce emissions into the environment—an engineer, George Best, was hired to deal with air and water pollution. Expenditures on health and environmental controls amounted to more than 10% of the total cost of the plant—an enormous sum at a time when DuPont, probably the most technologically advanced of the large chemical companies, had devoted less than 1% of its capital investment to environmental control. Tarr threw himself into the work; by the following April he fell ill from what was diagnosed as overwork and was under doctor's orders to limit his time in the plant to one hour per day. The new plant opened in 1951.[32]

Hueper in the meantime was back at work on the chromium problem. Over the objections of the Industrial Hygiene Division, NCI funded a study by Thomas Mancuso of the Ohio Health Department at a plant in that state. A series of publications describing that work appeared in 1951. A paper jointly authored by Mancuso and Hueper confirmed the earlier finding of elevated lung cancer rates and raised the question of whether the hazard was limited to the hexavalent form of the element, as generally believed, or extended to the trivalent form as well.[33]

A second paper drafted jointly by the two men followed a few months later, offering data that suggested a danger from trivalent chromium and raising the possibility that the cancer hazard extended to the population outside chromium plants. Hueper gave the manuscript of this yet-to-be-published paper to Tarr, who forwarded it to Lanza. Lanza sought deletion of the troublesome conclusions; when the request was refused, he complained to the Public Health Service. Seward Miller, the new chief of the Industrial

Hygiene Division, told Hueper to take his name off the paper; Hueper acceded but complained to Surgeon General Scheele. The surgeon general then shut down Hueper's entire program of activity outside the laboratory. Hueper was ordered to discontinue work on chromium, end his field work, and cease all contact with industry and with state and local health agencies. Michael Shimkin, who at the time of these events headed NCI's Laboratory of Experimental Oncology, later explained that Scheele thought action necessary "to preserve Federal programs in industrial hygiene more general than carcinogenic hazards." More likely, the express or implied threat was congressional action against *other* Public Health Service programs; Clarence Cannon, who in 1937 had redirected pesticide research funding from the FDA to the Industrial Hygiene Division, now chaired the House Appropriations Committee.[34]

Tarr's opinion of these events went unrecorded, if it was ever expressed in words at all. But on July 25—even before the order to stop work was passed down—he chose to tell Hueper what Lanza had done to him. When Tarr's activities are examined at a remove of 60 years, his role seems ambiguous; yet Hueper recounted in 1955 that he "had enjoyed the most harmonious cooperative relations" with Mutual Chemical during Tarr's lifetime.[35] Hueper's information was incomplete—he was unaware that Mutual had withheld from him in 1946 its knowledge of high cancer rates among its workers—but he knew much of what had happened, and he was parsimonious, to say the least, with praise for industrial managers. His positive evaluation of Tarr can hardly be dismissed.

Two years later, Hueper managed to resume research on chromium, establishing "informal" contact with Mutual and obtaining via Tarr blood samples from employees. Less than six months after it restarted, Hueper's chromium research was again shut down by his superiors. On Nov. 30, 1953, Hueper wrote to Tarr that he had been ordered to conclude his work within another six months. Tarr forwarded Hueper's letter to company president George Benington, remarking with evident irony that

> The fact that Doctor Hueper has been requested to discontinue his work within the next six months I presume has to do with the availability of funds.

Tarr went on to suggest that "our group might consider asking either Doctor Lanza or Doctor Baetjer to finish the work."[36] Anna Baetjer of Johns Hopkins University had already embarked on her studies of Mutual's employees, producing results that are now considered a scientific landmark. Where Hueper held a precautionary view of environmental hazards and urged action to protect human health without waiting for scientific certainty, Baetjer pursued a different approach. A recent admiring profile remarks that "her steadfast determination to accept only scientifically proven links between causes in the workplace and effects on worker health won her considerable influence...."[37]

Mutual Chemical was by now up for sale as a delayed consequence of the death of company president H. M. Kaufmann, following several years after his son and heir apparent had succumbed to Hodgkin's disease.[38] Tarr, in reply to a mid-1953 inquiry about safe levels from a German chromium producer, opined that the exposure limit in use was based entirely on acute illness and "we would not feel safe in assuming that the hazard with respect to lung cancer would be eliminated." But in a memorandum prepared a few months later for the prospective buyers, he put a more positive slant on what amounted to the same conclusion. On the subject of "control" (control of what is never stated, but lung cancer is obviously intended), he observed that while "It is not to be expected that reputable members of the medical profession...will undertake to guarantee that the measures adopted are certain to be adequate.....[nevertheless] we probably have solved our problem." The first evidence offered in support of this conclusion was this:[39]

> We have never had a case among our office, routine laboratory, research laboratory or engineering department employees, or among any employees who have not spent a major part of their working hours directly in the operations areas.

Sixteen months later Omar Tarr was dead of lung cancer.[40]

CHAPTER 7

Bad Air in Los Angeles

I just went into the office and said to Norman, "Something has to be done."

—Dorothy Buffum Chandler, recalling events of 1946[1]

S MOG WOULD COME TO define the essence of Los Angeles, but the city was a latecomer to it. The word was coined in London at the end of the nineteenth century to describe the mixture of coal smoke and fog that had plagued that city for centuries. It first gained wide circulation in 1905 when Henry des Voeux, an officer of the Coal Smoke Abatement Society, began to use it at scientific meetings.[2]

London smog was an object of fascination. A recent study explains artistic depictions of the murk—by painter Claude Monet and a multitude of writers—through the chemistry of coal tar, the source of the dye chemicals that were so important to the early study of environmental cancer. The yellow morning fog, Oscar Wilde's "ochre-coloured hay," was tinted by tars in the smoke of household coal furnaces that burned at low temperature. By afternoon, the dark smoke from hotter-burning industrial furnaces would turn the smog to brown or black.[3]

In America's industrial cities, coal smoke lacked London's complexity of color and chemistry, but it was no less an affliction. Municipal reformers of the New Deal era, like their predecessors a generation earlier, railed against the evils of bad air, but in most cities the agitation had little effect. Only St. Louis, at the end of the 1930s, made itself an exception.

The St. Louis achievements were captained by Raymond Tucker, a mechanical engineering professor from a local university. Tucker turned for advice to the Illinois State Geological Survey, where the problem was assigned to a team of researchers. Among them was a young coal geologist who was to become an important figure in air pollution control, Louis McCabe (1904–78). Born in Graphic, Arkansas, McCabe had worked for the state agency as an undergraduate at Northwestern University. After two years with a coal company, he rejoined the Survey as a coal geologist and earned a doctorate in 1937.[4]

Tucker and his advisors developed a new strategy for smoke control, focused on the fuel rather than the combustion process. After a severe pollution episode on November 28, 1939, regulations were put in place that allowed only high-grade coal to be burned. Boilers had to be modernized. With these rules strictly enforced, St. Louis by the winter of 1941 could boast of a 72% reduction in visible smoke. Other industrial cities such as Pittsburgh and Cincinnati would follow its lead after the war.[5]

Los Angeles in these years was a refuge from the bad air of northeastern industrial centers, famed for its healthy atmosphere. But soon a persistent haze emerged to scourge the City of Angels. By 1944 a word was borrowed from England, *smog*, that would be linked ever after with the city. As in London, the pall had a complexity that was perceived by artistically minded observers before it came to be understood by chemists. But what fouled the dry Southern California air was something new, more diverse in its sources than London's coal-darkened fogs while similarly pernicious in its effects. Blame for southern California's smog was at first laid exclusively on industry; a decade of research, some of it pioneering work in chemistry, showed automobile emissions to be a separate cause, harder to control and steadily growing in magnitude.

The first English-speaking settlers of Southern California thought of the region as a paradise of healthful air. Victims of tuberculosis from the smoky cities of the east fled westward to sanatoriums, which multiplied in Pasadena and nearby towns. But the area already suffered from episodes of air pollution. In 1868, when the population of Los Angeles was only 5,000, the air was filled for five days with thick smoke. Smoky days became more frequent when the area began to industrialize at the end of the nineteenth century. By 1903, a newspaper editorialist could describe the sky becoming so dark that a solar eclipse was thought to be under way.

The problem was the climate, at first thought so beneficent, and the geography. Los Angeles and its environs occupy a series of desert basins whose flat alluvial bottoms are separated by linear mountain ranges. Hot summers and calm winds favor the formation of temperature inversions, in which cooler air—sliding easily from the ocean into the flat-bottomed bowls of the coastal basins—is trapped beneath the hotter air above. The layer of cool air just above the ground can remain stagnant for days, polluted more and more as it collects smoke, tailpipe emissions, and whatever else human ingenuity contrives to send skyward.[6]

In the early years of the twentieth century, oil was discovered in the Los Angeles basin, and it was quickly put to use. Coal smoke, the bane of eastern cities, ceased to be a problem, and the region's air gained a respite of decades. But weather records show that visibility began to drop in 1939. In June and July of 1940, intervals of smoky air made eyes water and throats burn. The city health department was called in to investigate; while the cause could not be pinpointed, conditions were seen to be worst near certain industrial plants. These episodes recurred in the fall, and again the following summer. A four-hour attack on December 10, 1941, was the worst of the year.

Three days earlier, Pearl Harbor had been attacked. The Second World War brought a manufacturing boom to Los Angeles. The city's factories grew so fast that their consumption of electricity more than tripled from 1941 to 1943. And among the fastest-growing branches of industry were the makers of chemicals, rubber, and nonferrous metals, all heavy emitters of irritating fumes.

In July 1943, the plague of bad air returned with a new ferocity. The problem was blamed at first on automobile traffic caused by a streetcar strike, but the end of the strike on the 23rd brought no improvement. Complaints streamed into local governments; inspectors already searching for the causes of the air pollution problem were ordered to redouble their efforts. By the morning of the 26th, the air reeked as a thick cloud cut downtown visibility to three blocks. At noon, the siege lifted, and the afternoon sky was suddenly clear.

Attention soon fixed on a big chemical plant at the edge of downtown. With supplies of natural rubber cut off by Japanese war gains, the federal government had established a Rubber Reserve Corporation with the mission of creating almost instantly a synthetic rubber industry. An essential intermediate in rubber production was butadiene, manufactured from the petroleum feedstocks that were abundant in Los Angeles. Rubber Reserve had ordered the Southern California Gas Company to convert its Aliso Street gasworks, located close to downtown gas customers, to the production of that chemical. At a cost of $14 million, a hasty rebuilding was undertaken. The new plant, quickly thrown together, washed impurities from the product with water that was then sent into a cooling tower. There the impurities were liberated into the air, spreading noxious gases over a five-mile radius.

Controversy raged for three months as local politicians sent their constituents' complaints to Washington. Finally in October, the plant was shut down while pollution controls were installed. Production resumed just before Christmas, and with the butadiene plant's noxious fumes gone, the city's population thought the air pollution problem solved.

Health Department investigators knew better. They had measured air contamination upwind of the plant as bad as downwind, and they knew that irritation could set in at points far removed from each other. No one source could be at fault; the Board of Health was warned in May that a repeat of the previous summer's events was "inevitable." Efforts were made to educate the public, but they had limited effect. The false lesson from the butadiene episode would, for years to come, color Los Angeles' efforts to control its air pollution problems.[7]

In the summer of 1944 the noxious fumes returned as predicted, even worse than the year before. On some days, when inversions had low ceilings, the tops of buildings could be seen poking out of the miasma. Angry citizens again organized, circulating petitions that called for the control of fumes from a refinery and a chemical plant in El Segundo. On September 18, for the first time, the plague received the appellation of "smog" in the pages of the *Los Angeles Times*.

An unpaid commission, appointed the previous year, sprang to life and proposed ordinances for control of factory emissions. The rules were simple: only the visible opacity of smoke was limited. This was just a rough stab at a solution. Organic chemicals were already known to be at the core of the problem, but beyond the constraints imposed indirectly by the restriction on dark smoke, their emissions would not be limited. Still worse, the structure of California government made it hard to put even this straightforward scheme in place. Separate enactments were required from each of Los Angeles County's many cities and, for unincorporated areas, the county itself.

At this point there opened a cleavage between the city's civic leadership and its manufacturers which, for the next half-dozen years, would dominate the debate over the newly labeled affliction. Despite the mildness of the proposed regulations, the tentative effort at control quickly attracted the unfriendly attention of the Chamber of Commerce. After an initial plea to avoid "precipitate action" fell on deaf ears, the Chamber began to argue that industry had been unfairly singled out. Manufacturing interests made their usual threefold argument: more research was needed to understand the problem, controls should have a broader scope than industry alone, and a voluntary, cooperative approach would be more effective than regulation.[8]

Ordinances, weakened to meet the business objections, were adopted by Los Angeles City in November 1944, and by the county the following February. Each added a full-time air pollution control officer to the respective health department, with a mandate to investigate the causes of smog and make recommendations for further action. Emission limits, enforcement, and penalties were lacking. The county position was filled with a man of distinction, Isador Deutsch, a chemical engineer who had spent 20 years at Chicago's Department of Smoke Inspection and Abatement. Deutsch and his city counterpart, Harry Kunkel, quickly renewed the appeal for a smoke ordinance. Business opposition caused the legislation to linger through the summer, but both city and county finally enacted opacity limits in the fall.[9]

That research really was needed was apparent to Deutsch even before he arrived in May 1945. Inspectors went out to sources of visible smoke, because control of visible smoke was what was understood, and some genuine progress was made in reducing emissions from railroads, trucks, and factories. But in Los Angeles that was only a small part of the problem. The eye irritation that was so characteristic of the city's smog was absent in eastern cities. Indeed, tests soon showed that the city's dustfall was only 60% of Chicago's, and there was little correlation between visibility and the volume of particulates in the air. Nor was the culprit sulfur dioxide, the compound that had triggered so many smelter controversies during the preceding half century. Tests showed that Los Angeles smog contained less of the gas than Chicago air on sunny days.

Within months, Deutsch came to suspect that automobile exhaust was an important cause of smog. He did not think it the whole problem; oil refineries were obvious sources of compounds similar to what comes out of cars. But he did try, without being too open about his unproven theories, to get research started on the topic.[10]

The scientific establishment was skeptical of such speculation, and the public was downright hostile. For most Angelenos, the culprit was easy to identify: the factories that multiplied during and after the war. Indeed, the thinly staffed control agencies with their programs of voluntary cooperation were hard put to keep up with the continuing growth of the Southern California economy. Hopes for immediate success were dashed when the summer of 1946 was smoggier than ever. One month before the 1946 election, District Attorney Fred Howser, a candidate for statewide office, filed headline-making lawsuits against 13 industrial plants.[11]

Public opinion, in its single-minded assignment to industry of the blame for smog, was not completely off the mark. According to later estimates, emissions of organic gases from gasoline-powered vehicles in 1940 were

Smog shrouds Los Angeles City Hall, September 23, 1949.
At its top, the building rises above the temperature inversion
into clear air. (Courtesy of the University of Southern
California Digital Archive, Los Angeles Examiner Collection.)

580 tons per day, compared to 2,290 tons from stationary sources. As late as
1950, vehicles contributed 1,160 tons per day, while the stationary sources
were still slightly greater at 1,280 tons.

Defects in the control system were easy to see. Three small municipalities, Vernon, El Segundo, and Torrance, had been incorporated in the early
part of the century as low-tax havens for heavy industry. Vernon, where 10%
of the entire metropolitan area's factory workers labored but only 417 residents were housed, bore the words "Exclusively Industrial" on its city seal.
These cities harbored many of the most polluting factories, but they were
outside the jurisdiction of the county ordinance and refused to enact rules
of their own. To overcome the area's administrative fragmentation, a state

"The cop that cried." Irritating smog brings tears to the eyes of City Hall guard F. W. Styles, April 4, 1952. (Courtesy of the University of Southern California Digital Archive, Los Angeles Examiner Collection.)

law creating an air pollution district would be needed. In July 1946, the Los Angeles City Council called for action, and county officials soon began to draft a bill.[12]

With the smog worsening despite control efforts, demands for stronger remedies grew. Residents of Pasadena, whose economy still depended on tourism, were the first to organize; they were soon joined by one of the city's most influential men, Stephen Royce. Royce, the gregarious owner of the Huntington Hotel, was so upset by the smog's effect on the business of this luxury winter resort that he hired a full-time lobbyist at his own expense. The Pasadenans attracted support throughout the county.

The drive for air pollution legislation gained crucial support in the fall when the *Los Angeles Times*, the city's leading newspaper, launched a full-throated campaign on the issue. The decision to set the family-owned *Times* on this course was made by its publisher, Norman Chandler, at the urging of his opinionated wife Dorothy Buffum Chandler, known as Buffie, the region's leading patron of the arts. As she recalled the story years afterward, she was driving over the mountains into the city one day. The sight of an ugly layer of haze covering the valley moved her to immediate action. "I just went into the office and said to Norman, 'Something has to be done.'" This recollection may be somewhat dramatized—Norman, in a 1957 interview, placed the crucial conversation in their car—but it conveys the essence of what happened. The campaign was launched with the assignment of noted local columnist Ed Ainsworth as a full-time smog editor. By October Ainsworth was writing feature stories about the work of control agencies almost daily, and citizen complaints began to pour into the newspaper.

The paper's next move was to hire as a consultant Raymond Tucker, the manager of St. Louis's uniquely successful prewar smoke control program. Tucker's two-week visit in December was heralded by a storm of publicity, climaxing on the day of his arrival with the entire editorial page of the *Times* devoted to air pollution. Tucker's time in Los Angeles was a whirlwind of meetings and visits, missing only one thing—smog; the air was clear for the entire period. On his return to St. Louis, Tucker quickly prepared a report, published by the *Times* on the front page of the January 19 Sunday paper and reprinted as a free pamphlet. The Tucker report broke no new scientific ground, it downplayed automobiles as a source of pollution, and its 23 recommendations drew heavily on the work of local experts; yet it was crucially important in making clear that halfway measures would not work and an unprecedented control effort would be required.

Meanwhile, Chandler had organized a prestigious citizens committee to back up his efforts. Chaired by William Jeffers, former president of the Union Pacific Railroad and of the Rubber Reserve Corporation, the group also included the Nobel Prize–winning physicist Robert Milliken, who had been installed as president of Caltech by Chandler's father. Stephen Royce's lobbyist, still on the hotelier's payroll, became its executive secretary. The committee served mainly as a vehicle to convince business interests not to block creation of the new pollution control agency. A first hurdle was overcome in April when the City of Los Angeles and other municipalities, which had initially sought to appoint the majority of the new district's board, agreed to cede control to the county supervisors.

Opposition remained from influential railroad, lumber, and oil-refining interests. The railroads were troubled by the prospect of a ban on old oil-burning freight engines, while refiners objected to the requirement of a license for construction of new or modified plants. Jeffers visited Sacramento in late April and won over the railroads with a promise that enforcement would be reasonable; the bill was thereupon approved by a near-unanimous Assembly. The oil and lumber industries then moved to bottle up the bill in a Senate committee. When a threat of bad press failed to dissuade the refiners' lobbyist, Jeffers was immediately informed. He turned to the *Times*, and a day later a stinging editorial appeared, lambasting the "selfish opposition of major oil and lumber companies" and concluding with an appeal for citizens to contact their state senators. Royce, acquainted with several oil company presidents who were frequent guests in his hotel, arranged for Jeffers and other control advocates to meet with the industry on May 20. The refiners, under heavy pressure from publicity and bombarded with complaints within their own social circles, capitulated. They were followed a day later by the lumber interests. The refiners' immediate reward was praise from the *Times* for past work on pollution control, accompanied by a bland statement from Jeffers that the reports of industry opposition had been "inaccurate." On June 2, the bill passed the Senate 29–0, and it was soon signed into law by Governor Earl Warren.[13]

The new agency required a new leader with a high profile. After Raymond Tucker declined the post, the county undertook a national search. Their selection—undoubtedly on Tucker's recommendation—was Louis McCabe. Following his work on air pollution in Illinois during the 1930s, McCabe had joined the Bureau of Mines during the war. By now he had risen to head the Bureau's coal division. He was given a two-year leave from the federal agency to take the Los Angeles post.[14]

Louis McCabe, director of the Air Pollution Control District, reports to the Los Angeles City Council, November 26, 1948. Mayor Fletcher Bowron and Council President Harold Henry are on the left. (Courtesy of the University of Southern California Digital Archive, Los Angeles Examiner Collection.)

McCabe, like Tucker and most other pollution specialists from outside the city, believed that smog was essentially a problem of industry. Although Deutsch's theories about automobiles were winning adherents among local specialists, McCabe at first discounted them and did not pursue research on the subject. The newly appointed smog chieftain launched a vigorous attack on the county's polluting factories. Rather than wait for definitive identification of the offending chemical agent, he set out to control the pollutants he knew about. Emission limits were proposed for lead, zinc, sulfur dioxide, and particulates. Twenty-two inspectors were hired, and an engineering staff of 12 was established to offer advice on cleanup methods and pursue research.

McCabe soon zeroed in on sulfur emissions from the city's refineries as a likely cause for the irritating smog. The danger to humans and to plants from sulfur dioxide was well known from decades of smelter studies, and the refineries' practice of burning waste gases released large amounts of the substance. The refineries fought back with the usual weapon—a call for research rather than action. The Western Oil and Gas Association had already begun their own research program at Stanford Research Institute, but their objections did not succeed in this case. The district gathered data that showed the extent of the refinery sulfur emissions, and McCabe stood his ground at a showdown with the oil industry in the presence of the county's highest officials. By February, 1949, the first sulfur recovery plant opened; cleanup was under way at the refineries.[15]

Louis McCabe soon returned to the Bureau of Mines in Washington, his work of setting up the new organization accomplished, but resistance to air pollution control continued. In the fall of 1949, Assemblyman Randal Dickey, flush with success after enacting the deregulation of liquid industrial wastes earlier in the year (see chapter 10), introduced a bill that would transfer air pollution control to a new state agency. This was seen instantly as an attack on pollution control, and the county succeeded in fending it off.[16] But soon more difficult obstacles were to be confronted, as control of industrial emissions failed to yield the hoped-for result of cleaner air, and new scientific discoveries clarified the growing importance of automobiles in creating the smog problem.

————

In 1944, farmers began reporting damage to vegetable crops in fields located between Los Angeles and Long Beach. The damage worsened each year, and by 1949 had spread to the north, south, and east, beyond the confines of the basin. By then the damage had mounted to a half-million dollars per year, with entire fields of spinach destroyed.

Experiments proved the blight was caused by air pollution, but after initial suggestions that sulfur dioxide might be the cause proved unfounded, agronomists were unable to identify the active chemical agent. The Chamber of Commerce's scientific committee, chaired by Robert Vivian of Caltech, decided to drop the subject until some new ideas came along. Vivian then received a short note from a Caltech chemist who specialized in analyzing flavors and smells of foods, Arie Haagen-Smit. "Why don't they analyze the air?" asked Haagen-Smit, who in late 1948 started to work informally on the problem in his own laboratory.

Haagen-Smit began by condensing air samples to test the theory favored by control district chemists, that the culprits were chlorinated hydrocarbon compounds. He did not find what he was looking for, but he now had liquid samples he could analyze in his laboratory. By February he was able to report a significant discovery. He had found organic peroxides, chemicals that could be traced back only to petroleum. This suggested that chemical reactions were occurring in the air, converting the compounds released into the air into something different that made up the smog. The pattern of morning smog formation, as seen from the hills, seemed to back this idea up. Other supporting evidence started to accumulate: correlations between eye irritation and sunlight, concentrations of certain compounds that far exceeded what could be explained by known sources, and mounting evidence that Los Angeles air had unusually high levels of ozone.

Haagen-Smit then conducted a crucial experiment. He combined ozone and gasoline in his laboratory and produced what looked and smelled like smog. Plants exposed to this brew developed all the usual symptoms of smog damage. Further experiments showed that ozone could be produced by shining ultraviolet radiation, the component of sunlight most likely to stimulate chemical reactions, on a mixture of the olefins that are found in gasoline and the nitrogen dioxide that is formed in car engines. Haagen-Smit took the year of 1950–51 off from Caltech to pursue his research full-time. He was joined by Bureau of Mines scientists lent by Louis McCabe.[17]

The photochemical smog theory encountered stiff resistance. Haagen-Smit's first papers were rejected by scientific journals. He came under heavy fire from the Western Oil and Gas Association, whose refineries shared the blame with automobiles as heavy emitters of hydrocarbons. The refiners' well-funded research program at Stanford Research Institute suggested one alternative theory after another, all tending to exonerate the refineries and all backed up by a lavish public relations effort. When the Second National Air Pollution conference was held at the Huntington Hotel in 1952, Haagen-Smit was able to present a complete theory. He identified the key chemical reactions, provided experimental confirmation, and correlated them with crop damage. To forestall any attempt by the refiners to have their own research overshadow

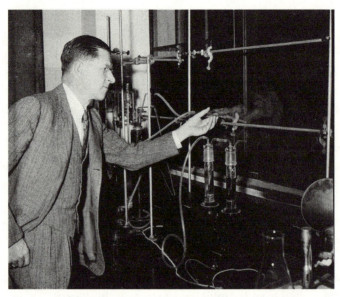

Arie Haagen-Smit demonstrates the manufacture of artificial smog in his Caltech laboratory, November 17, 1950. The tubes bring together ozone with hydrocarbons, triggering a smog-creating chemical reaction. (Courtesy of the University of Southern California Digital Archive, Los Angeles Examiner Collection.)

this presentation, the control district arranged to have Haagen-Smit's work written up in the *Times* just before the conference began.

Meanwhile, the control district began enforcement activities against the oil industry. Inspectors in 1951 examined all of the county's refineries and found that they allowed 800 tons per year of gasoline and other hydrocarbons to evaporate into the air. Within three years, losses were cut to 220 tons. By 1953, pressure was increasing on both the oil and automobile industries, based on a theory that the industries still did not accept but could not disprove. At this point a new research body was established, the Southern California Air Pollution Foundation, ostensibly a broad-based community effort but actually sustained by the refiners. Alongside an unprecedentedly well-funded research effort, the foundation organized a campaign to persuade the public of the dangers of overly hasty action.

Up to this point, the industry researchers at Stanford Research Institute had been unable to duplicate Haagen-Smit's results. But after the Caltech chemist personally demonstrated his technique to scientists from industry-funded institutes, they were replicated in 1955 at Illinois Institute of

Technology, the Franklin Institute in Philadelphia, and finally at SRI. The chemical perpetrators of air pollution in Los Angeles had been found. Scientific controversy around the cause of Los Angeles' smog was at an end; the debate over what to do about it would continue for decades to come.[18]

With smog, as with other environmental scourges of this and other eras, industry's recurrent refrain was a call for research before regulation. In this instance, research did indeed win manufacturers partial exoneration. But it was the regulators, not industry itself, who made the crucial discovery about California smog—it is created by chemical reactions in the sunlit atmosphere. Industry's own research began in earnest only after the discovery was made, and was at first devoted to casting doubt on it. Only later, when no doubt remained, did industry redirect its resources. Learning something new about the still-growing smog problem, instead of just raising questions about the findings of others, was their last resort.

| Donora's Strangler Smog

We have to get over the idea that smog is just a nuisance. There is no condition confronting us that is more terrifying.

—James Townsend, Nov. 18, 1948

The starting point of the whole problem, therefore, is more research.

—James Townsend, Oct. 28, 1949[1]

DONORA, PENNSYLVANIA, IN 1948 was a hardworking factory town nestled deep in the valley of the Monongahela River, not far north of the West Virginia state line. With its steel and zinc plants, both owned by U.S. Steel, Donora was part of an industrial behemoth that could take much of the credit for winning the great war that had just ended—traveling down the river to Pittsburgh, one would rarely be out of sight of a mill.

Smoke from the factories was a constant element of the community's life. The aged zinc plant belched out prodigious amounts of fume and dust full of sulfur, zinc, lead, cadmium, and arsenic. Control technology existed—these emissions could be trapped in baghouses and electrostatic precipitators—but U.S. Steel, like its competitors, only installed these devices where it was profitable to do so. Thus the plant used the sulfur from its roasters to make acid and collected metallic fume from ore sintering. The sale of these by-products was vital to the economics of zinc making, a business whose personnel, technology, and products defined it as part of the chemical industry. The smoke from the furnaces that smelted the zinc had no value, so it was left to rise unhindered into the air.[2]

Sitting on a horseshoe bend of the river, with the ground sloping steeply up from the river valley, Donora is sheltered from the wind. As in Los Angeles, temperature inversions trap smoke in the bowl formed by the valley walls. The wetter eastern climate made dark smoky fogs a familiar occurrence, especially in the humid weather of autumn.

An especially severe inversion formed above Donora toward the end of October 1948. On Tuesday, the 26th, the weather began to close in, and the morning fog took longer than usual to burn off. The next day was worse, and on Thursday the town was closed in. The renowned medical writer Berton Roueché portrayed the scene:

> ...it had stiffened adhesively into a motionless clot of smoke. That afternoon, it was just possible to see across the street, and, except for the stacks, the mills had vanished. The air began to have a sickening smell, almost a taste.... but no one was much concerned. The smell of sulphur dioxide, a scratchy gas given off by burning coal and melting ore, is a normal concomitant of any durable fog in Donora. This time, it merely seemed more penetrating than usual.

On Friday morning, the telephones of the town's eight doctors began to ring. By evening, the phones were ringing incessantly. The doctors were up all night, as was the fire department, bringing oxygen to the sick. The first death occurred on Friday afternoon, and by Saturday morning the town's leading undertaker had collected nine corpses.[3]

At noon Friday, the superintendant of the zinc works made a regular check of the dispersion of smoke from its stack and saw no reason to bank the plant's ovens. At lunch, though, one of the plant's managers was reminded of what had happened in the Meuse valley of Belgium in December 1930. Five days of smoky fog in an industrial valley had killed more than 60 people and made hundreds gravely ill. The comparison was well chosen. In the Meuse, as in Donora, illness struck mostly the elderly, with asthma-like symptoms aggravated by cardiovascular difficulties. And in both instances, the population laid immediate blame on the zinc factories.[4]

By Saturday, residents were fleeing Donora, as medical assistance arrived from outside. In the morning, and again in the afternoon, the zinc plant checked its stack and decided to keep running. Newspaper and wire-service reporters had arrived by morning; in the evening, a national radio broadcast

"Midnight at noon." The smog-darkened city of Donora on October 29, 1948. (Copyright *Pittsburgh Post-Gazette,* 2009, all rights reserved. Reprinted with permission.)

by gossip columnist Walter Winchell drew wide attention to the disaster. So many people were trying to leave town on smog-shrouded roads that evacuation became nearly impossible.

Around 3:00 A.M. Sunday, U.S. Steel's general counsel, Roger Blough, called from Delaware and told the manager of the Donora zinc mill to stop production. At 6:00 A.M., the smoke-generating feed of ore to the furnaces was halted. Later that morning, it began to rain, and the smoke dissipated. Nineteen of the town's 14,000 residents were dead of asphyxiation and 45, one of whom would soon succumb, were in hospitals. Nearly half the town's population was ill.[5]

Even before the smoke cleared, investigations began into the causes of the mysterious disaster. Few, at this point, doubted that the zinc works were the source of the poisonous emissions. Pittsburgh's health director, Hope Alexander, told the Associated Press on Sunday that most of the Donora deaths had occurred within two or three blocks of the zinc smelter. The next night he was joined in blaming the zinc mill by local doctor William Rongaus and Frank Burke, national safety director of the United Steelworkers of America, which represented the workers at both Donora plants. Unidentified plant

The Donora zinc mill, probable source of the toxic air pollution that enveloped the town in 1948. (Courtesy of the National Library of Medicine, History of Medicine Division.)

spokesmen denied responsibility to the *New York Times*, but the Associated Press was told that the "objectionable fumes" that had caused the shutdown were zinc oxide. Worry about this compound was nothing new for air pollution specialists; Louis McCabe's first set of Los Angeles air pollution regulations, proposed ten months earlier, had listed zinc oxide among the three compounds to be controlled.[6]

As early as Saturday morning, the local Board of Health and the Pennsylvania Health Department asked for help from the U.S. Public Health Service. They were refused. The state renewed the request the next morning, and was refused again. The first investigators to arrive were from the Industrial Hygiene Foundation in Pittsburgh, summoned by U.S. Steel. The foundation's team reached Donora shortly after 6:00 A.M. on Sunday, followed later in the morning by industrial hygienists from the Pennsylvania Department of Health. Scientists from the Bureau of Mines, the one federal agency experienced with air pollution, arrived on Monday. At that evening's meeting, Burke of the Steelworkers was joined by the union's local attorney, Jason Richardson, in calling again for an investigation by the Public Health Service. The union distrusted both the industry foundation and the Republican-controlled state government.[7]

The sun at last shone brightly on Donora on Tuesday morning as eight of the twenty victims were put to rest.[8] Tuesday, November 2, 1948, was election day. The 1948 election was perhaps the most economically polarized in the history of the United States; labor was the core of President Harry Truman's reelection campaign, and the Steelworkers were the backbone of the more activist of the two labor federations, the CIO. Donora, a solid union town, gave Truman 2,798 votes, against 1,077 for the Republican ticket of Thomas Dewey and Earl Warren.[9] There was little hope as the town cast its votes that labor's candidate would prevail; Dewey was the overwhelming favorite. But Donora and the nation awoke Wednesday to the astonishing news that Truman was ahead, and by midmorning the Democrat was a clear victor. Behind him, the Democrats seized control of both houses of Congress.

This political earthquake shook the complex landscape of competing economic and bureaucratic interests that had assembled in Donora, setting the investigation off on a zigzag course. On Tuesday, the Industrial Hygiene Division's assistant chief, John Bloomfield, had once again rebuffed the calls for a Public Health Service investigation, ascribing the disaster to an "atmospheric freak." On Wednesday, with the electoral winds behind him, the Steelworkers union's number two, International Secretary-Treasurer David MacDonald, wrote to Pennsylvania's governor to renew the appeal for a federal investigation. The Steelworkers were now heard in Washington. The next morning, a government industrial hygienist arrived in Donora. Federal and state scientists recommended a plan for the zinc works to reopen on Monday, with air quality monitored by a network of permanent samplers. By Friday, the Bureau of Mines had agreed to lend an experienced air pollution researcher, Helmuth Schrenk, to the Industrial Hygiene Division to serve as study director.[10]

Almost certainly, this sudden reversal resulted from direct intervention by Steelworkers president Philip Murray with President Truman. The amplitude, rapidity, and breadth of the federal response—not only did the Public Health Service abruptly reverse course, but the Bureau of Mines, part of the Interior Department, moved with breathtaking speed to help another agency—are hard to explain otherwise. Murray, as president of the CIO, was in regular contact with Truman at election time. Most likely it was on the morning of Election Day, while the president was away from Washington, that the two of them discussed the issue.[11] Truman did not return from the campaign trail until Friday, November 5; the president's appointments calendar shows that just three hours after his arrival at the White House that day he met consecutively with Oscar Ewing, who oversaw the Public Health

Service, and Murray. Donora, it appears, was no small matter; at this date, six months into the Berlin airlift, the cold war was well under way, yet the President's meetings with Ewing and Murray came before any appointments dealing with military or diplomatic topics.[12]

Within days, the Industrial Hygiene Division's U-turn was complete. Both Bloomfield and James Townsend, the Division's chief, spent Nov. 17 in Donora planning the investigation. They outlined their plans at an evening public meeting. Steelworkers union lawyer Jason Richardson reported back to Frank Burke on this gathering, adding a note of caution:

> I feel sure that the Public Health Service force will be impartial and try to do a good job, but of course in this, as in any other field, influence can be brought to bear. I would recommend that some constant contact be kept with either Doctor Townsend or Doctor Bloomfield to see that the impartiality of their investigation is maintained, and that we be afforded some access to their findings and reports.[13]

The next day, Townsend and Bloomfield went on to Pittsburgh for a meeting of the Industrial Hygiene Foundation. There Townsend declared that the Donora smog was

> ... a problem that transcends Donora and is nation-wide in scope. ... We have to get over the idea that smog is just a nuisance. There is no condition confronting us that is more terrifying.

The sudden expansion of the Industrial Hygiene Division's mandate from the factory to the outside air went beyond Donora. While in Pittsburgh for the IHF meeting, Townsend submitted a plan for the forthcoming investigation of lung cancer at chromium plants to representatives of that industry. The scope of the study was no longer to be limited to conditions within the plant, but also included research into chromium exposures in neighboring communities.[14]

Industry did not let these events pass in silence. Both zinc and chromium manufacturers insisted that their air emissions were not a danger. The U.S. Steel subsidiary that operated the zinc smelter in Donora issued a public statement on Nov. 16 asserting that it was "certain" that the zinc works were not responsible for the smog deaths. At the same time, the company launched a series of scientific studies aimed at preparing a defense against expected lawsuits. For this purpose, Robert Kehoe's Kettering Institute was brought in to replace the Industrial Hygiene Foundation, whose policies forbade court testimony. Kettering began work the first week of December, carrying out its own toxicological, meteorological, and industrial hygiene surveys.[15]

The Steelworkers' fear that "influence can be brought to bear" proved to be well founded. The Industrial Hygiene Division soon reversed course a second time.

Within weeks, air pollution was eliminated from the planned chromium study. This decision was made on December 29, when Omar Tarr and Mutual Chemical's chief counsel Theodore Waters met with Townsend and Bloomfield to discuss the plan they had been given on November 18. Tarr and Waters asked for deletion of all consideration of off-site air pollution from the plan; the request was granted on the spot, and the scope of the study was limited to the industry's workers.[16]

Community air pollution could not, of course, be omitted from the Donora study. The Industrial Hygiene Division lacked air pollution expertise and required outside help. An obvious place to turn was the Bureau of Mines, whose mission had included air pollution since its founding. Helmuth Schrenk, a long-time collaborator of Royd Sayers in industrial hygiene research, was borrowed from that agency to lead the investigation. But with Sayers having been pushed out of the Mines directorship a year earlier, the Bureau could not be trusted to offer the advice that the Division wanted to hear. As early as the 1930s, the Bureau's safety division under Daniel Harrington had quietly disagreed with the Public Health Service unit's proclivity to play down the hazards of breathing dangerous materials.

Instead, the Industrial Hygiene Division allowed its work to come under the strong influence of Kehoe and his assistants at Kettering. Kettering scientists were allowed to join in planning sessions, where they pointed the study in directions congenial to U.S. Steel. Rather than collecting air samples near the zinc plant, a monitoring network was spread throughout the community. This would help to understand the weather conditions that had triggered the disaster, but was of little use to determine the amounts of contaminants carried by the deadly smog or to pin down their source. The Bureau of Mines kept its distance from this misdirected sampling plan and resisted requests for help. Asked to analyze 100 air samples, it negotiated the number down to 25.

The heart of the Donora investigation was histological analysis of the victims' lung tissues. This was the primary means of determining the cause of death, and therefore the key to finding the cause of the disaster. Instead of calling on Public Health Service resources at the National Institutes of Health, the Industrial Hygiene Division contracted this task out to Arthur Vorwald of the Saranac Laboratory, who had just agreed to revise his asbestos report to remove all mention of cancer (see chapter 6 above). At Donora, Vorwald shared unpublished data from his government-funded tissue analyses with U.S. Steel's litigation experts and allowed them to join with him in carrying out the autopsies of two victims. Vorwald was then entrusted with the crucial responsibility of synthesizing the entire government investigation to identify the toxic agent in the smog.[17]

While inviting Kettering's industry-sponsored researchers into the study, the PHS kept independent experts—the most credible of whom was Clarence Mills, a colleague of Kehoe as head of the Department of Experimental Medicine at the University of Cincinnati's medical school and a leading campaigner for air pollution control—at a distance. On the Saturday following the disaster, the Steelworkers had offered the town $10,000 for independent studies. The town used this money to implement Mills' suggestion of a community-wide health survey; a door-to-door canvass began on December 8. On December 24, while the canvass was still under way, Mills issued a report warning that the disaster could have been much worse:

> A slightly higher poison concentration in the air or a few hours longer time and the whole community might have been left almost devoid of life.

With local opinion strongly influenced by the town's dominant employer, U.S. Steel, such opinions were uncongenial. Adding emissions controls would cost as much as a new zinc plant, according to the plant manager, and the local union leaders who dominated the municipal government were loath to endanger the employment that sustained the town. Mills' assertions were condemned by community leaders. With the funds from the Steelworkers insufficient in any case to run an independent investigation, completion of the canvass was turned over to the Industrial Hygiene Division.[18]

As in Los Angeles, a local newspaper made air pollution its cause. When it became clear in the early months of 1949 that the Industrial Hygiene Division's investigation would avoid focusing on the zinc mill, sharply critical editorials started to appear in the *Monessen Daily Independent*, published in a mill town just upriver from Donora. Residents organized to seek enactment of a municipal smoke control ordinance, gaining the support of Pittsburgh Mayor David Lawrence. Victims soon began to file lawsuits against the zinc mill.[19]

After repeated delays, the Industrial Hygiene Division's "Preliminary Report"—the only report it ever issued—was released by Ewing and Scheele at a Washington press conference on October 13, 1949. The study was notably limited in scope. No effort was made to determine which of the city's factories had caused the health problems. Emissions from the zinc and steel mills were measured, but the report did not try to calculate the resulting concentrations of pollutants in the air. Having subcontracted the crucial task of identifying the chemical substance responsible for the illnesses to the industry-friendly Saranac Laboratory, the Division declared that no such identification was possible. In a return to the stance it had taken before the upset presidential election results became known, blame for the disaster was placed on rare weather conditions rather than any special danger from Donora's industries.

When he delivered a paper on the investigation October 28, James Townsend showed the distance he had traveled since the post-election days when smog was a terrifying condition of nationwide scope. The odds were now "extremely long" against recurrence, and "The starting point of the whole problem...is more research." This call for more research rings hollow indeed, coming just ten months after Townsend himself had quashed research on air pollution around chromium plants.[20]

Criticism of the report was immediate. The Steelworkers' Frank Burke crashed Ewing's press conference, denouncing the study for wrongly directing attention toward domestic heating and railroads rather than industrial sources. Clarence Mills followed in *Science* magazine with a scathing critique of the Public Health Service's unwillingness to draw conclusions from the data it had gathered. Had the concentrations of pollutants in Donora's air been calculated, he charged, the report would have found that levels of nitrogen oxides, zinc, and carbon monoxide were far above what was considered safe in factories.[21]

By March 1950, claims in lawsuits totaled $4.5 million. Litigation against the Donora zinc mill was not a novelty; local farmers had pursued long-running claims for similar sums in the '20s and '30s. The Industrial Hygiene Division had aided U.S. Steel's defense against the early suits by lending the services of John Bloomfield, who was now the deputy division chief, and once again the government came to the aid of the steel company. Unpublished data were supplied to Robert Kehoe's team of defense experts, which was bolstered by the hiring of such men as Anthony Lanza, M.I.T. meteorologist Hurd Willett, and Philip Drinker of Harvard Medical School. Helmuth Schrenk, who upon completion of the government study had left the Bureau of Mines to take a position with the Industrial Hygiene Foundation, also assisted the defense team.

The Kettering team undertook their own investigation, including medical examinations of more than 100 plaintiffs and air sampling inside and outside the plant. Working closely with defense attorneys, Kehoe prepared a detailed report arguing that the disaster was an act of God caused by unusual weather conditions. The existence of this document became known through a remark Kehoe made at a scientific meeting, but the plaintiffs' demand to see it was rejected by the courts on the grounds that it was expert opinion rather than fact. Denied access to this crucial information, the plaintiffs settled for $256,000 in 1951. The Donora air pollution disaster passed into history, the chemicals that poisoned so many remaining unidentified even though the evidence pointed plainly to their source.[22]

The Donora investigation had been neutered, but the wider implications of the disaster still could not be ignored. It fell initially to Anthony Lanza, who was to help the Donora defense prepare its cross-examination of Clarence Mills,[23] to present the lessons of the pollution episode as industry wanted them to be learned. Lanza's reassuring conclusions were laid out in a talk given May 24, 1949, to the national association of air pollution specialists and reprinted as a pamphlet:

> The common industrial poisons do not affect the general population as a result of atmospheric pollution....On rare occasions mother nature has played us false, and we have had the chain of circumstances that contributed such episodes as Donora...
>
> ...the greatest amount of pollution is in Pittsburgh. However, the incidence rate for lung and respiratory cancer in males is lower in Pittsburgh...[24]

The Bureau of Mines drew a different lesson from the disaster. Louis McCabe came back to Washington in mid-1949 to head its air and water pollution programs, and under his leadership the Bureau called for a federal air pollution program that went beyond research to control. Like the mine safety expert Daniel Harrington, who had just retired, McCabe was careful to avoid open disputes. He hedged his criticisms of industrial polluters with encomiums to the goodwill of the responsible majority of industry. But the critical tone of his public statements was unmistakable.

McCabe picked a high-profile occasion to speak out. He returned to Los Angeles in November for what was billed as the First National Air Pollution Symposium, held at Stephen Royce's Huntington Hotel in Pasadena. He gave a luncheon address that was quoted at length in newspapers across the country. Without identifying the objects of his criticism by name, McCabe heaped scorn on the arguments that Lanza had made six months earlier:

> In a Washington release in October is the statement that "smoke of itself is not a bad pollutant, except under unusual atmospheric conditions....[and it] is objectionable principally because of psychological effect...." Have we reached the state where we can gild refined gold or paint the lily?

...Unfortunately, the interpretation of health statistics is not reserved to the competent and the responsible, so we find growing out of the same body of vital data a catastrophe folklore on the one hand and a superman folklore on the other.... The claim is made that atmospheric pollution is the cause of respiratory cancer, and a layman claims that this is not true; witness—St. Louis, which passed a smoke-abatement law in 1940, has twice as many deaths from respiratory cancer as Pittsburgh...

"Why," he asked, "have we generally failed in our efforts to control air pollution?" McCabe offered three answers: "an incorrect estimate of the kind and size of the job"; "great waste of effort and misguided enthusiasm on the part of the public"; and finally

...industry believed that air-pollution control cost too much. Smoke and dusts were the wages of a prosperous industrial community, and the public generally shared this view or were forced by time and circumstance to accept it. There were "cooperative" programs with the dual objectives of delay and defeat. Engineers were assigned to write diverting papers on the minutiae of the problem, and the trade journals editorialized on the unreasonableness of "do-gooders." These tactics haunt the sincere efforts of progressive industry today.[25]

———

Why is it that scientists failed to unravel the mechanism of smog in Donora as they did in Los Angeles? Lack of resources is not the explanation; Donora received an unprecedented commitment of the federal government's resources at the direction of the president himself, while scientists interested in California smog often struggled for funding. And, while the difficulty of an unsolved riddle can never be known with certainty, there is every reason to think that the scientific questions posed in Donora were easier to answer. Researchers were able to establish that the problems of both areas were due to the combined effects of at least two distinct pollutants, but in Donora the jigsaw puzzle had many fewer pieces to put together. Geographic evidence there pointed directly to the source of the problem, while the spatial pattern of Los Angeles smog offered only vague clues. And Donora's smog, a single episode of a few days, clearly had a single explanation, while Los Angeles had recurrent problems whose nature and causes shifted over time.

The real difference between Los Angeles and Donora was economic. One was a major metropolis that controlled its own fate. The other was a factory town with an economy governed by a distant corporation. The fight against California smog began in earnest only when local elites, suffering along with everyone else, joined in. Buffie Chandler and Stephen Royce, who got things started in Los Angeles, might have stepped out of the pages of *The Official Preppie Handbook*; such people were not to be found in a Pennsylvania mill town. The *Monessen Daily Independent* was not the *Los Angeles Times*, and the men who ran U.S. Steel lived far away. Even in the months following the 1948 election, arguably the moment when American workers had more political power than ever before or since, inequalities of class weighed heavily on Donora's victims.

CHAPTER 9 | A New Deal for Clean Water

"The statute provisions are ample to protect the Schuylkill River from pollution; the difficulty lies in their enforcement, and they have remained almost entirely without life upon the statute books—an attempt was made to increase the penalties but the manufacturing interests kept it from passing..." The significant thing about this statement is that it was made in 1886 and referred to conditions of 1828 and 1832.

—Max Trumper, 1936[1]

I N WATER POLLUTION CONTROL as well as with mine safety and pesticides, the onset of the New Deal brought a return of government activism. But progress came only slowly and haltingly.

In the first Roosevelt administration, accomplishments in cleaning up streams came largely as by-products of programs with other goals. The Public Works Administration, aiming to fight unemployment, invested heavily in sewage treatment plants. The Social Security Act triggered a rapid expansion of public health agencies, enabling the Public Health Service to expand its research on stream pollution and undertake a program of acid mine drainage control. The National Resources Committee, established in 1935, carried out the first comprehensive analyses of water pollution as a national problem.[2]

While the new government agencies pursued this relatively uncontroversial work, pressure was building on the outside for an attack on industrial pollution. Proponents of control were gaining in numbers and influence as fishing, hunting, and other types of recreation grew in popularity. The Izaak Walton League, formed in 1922 to represent fishermen, recruited more than 100,000 members during the 1920s and developed a sophisticated lobbying operation. After successfully campaigning for creation of a game and fish preserve along the upper Mississippi, the group in 1926 decided to make control of water pollution its top priority.[3]

Clean water advocates had only scorn for the existing system of state water pollution control and demanded federal intervention. State capitals had long been under the sway of business interests, and the threat of moving operations to another jurisdiction was a potent counter to unwelcome oversight. A leader of the Izaak Walton League, Kenneth Reid, used his experience as a Pennsylvania fish commissioner to illustrate this point:

> The Pennsylvania State Legislature would not pass anything in this connection because it might penalize Pennsylvania industry in competition with New York and West Virginia and other States. The funny thing about it is that the same fight is going on in the legislatures of the other States, and they do not want to do anything until Pennsylvania does something.... As a result no State has done anything.[4]

In practice, the ability of large corporations to relocate tied the hands of state regulators. Louis Brandeis, the jurist who inspired many New Dealers, famously described the states as the laboratories of democracy. But—as a young Massachusetts legislator named Barney Frank would remark years later—Justice Brandeis's laboratories have a problem: you can't run a good experiment if the rats are free to choose their own cages.

The clean water drive was led in Congress by Senator Augustine Lonergan of Connecticut. In December 1934, Lonergan joined with the secretary of war, overseer of the Corps of Engineers, to sponsor a water pollution conference attended by government agencies and outside interest groups. One year later, after a conference subcommittee set up to make legislative recommendations had deadlocked, Lonergan gave a nationwide radio address on water pollution. A month afterwards his bill to establish a system of federal regulation, based largely on Izaak Walton League recommendations, was introduced. Hearings before the Senate Commerce Committee began on February 26.[5]

Industry was quick to react. The American Petroleum Institute, well equipped for battle after its years of struggle over oil pollution, took the lead in the legislative arena. The core of the lobbying strategy devised by the oilmen was undeviating opposition to any federal regulatory authority. An internal industry report explained that

Senator Augustine Lonergan, chief congressional sponsor of water pollution control legislation in the 1930s. (Courtesy of the Library of Congress, Harris & Ewing Collection.)

the only thing which practical politicians...recognize is an organized opposition capable of influencing votes. If the industry proceeds on that basis—plays poker rather than throwing down its cards in advance—it may again, as it did in 1924, postpone drastic anti-pollution legislation.[6]

As in the 1920s fight over oil, industrial interests accepted as inevitable that federal legislation of some sort would pass and sought to restrict its scope so as to make it ineffectual. Business interests lined up behind a bill submitted by Senator Alben Barkley of Kentucky which proposed a program of research, financial assistance to state agencies, and encouragement of voluntary action. Their lobbying was backed up with a communications strategy. With the depredations of the Standard Oil trust still in recent memory, the heirs of the monopoly were to stay in the background. The oil industry report recommended that "public statements should be confined to groups of producers—preferably independent organizations which are free of the political liabilities that some of the industry's major units have inherited from a fairly recent past." Supporters of the Barkley bill were recruited among engineers and public health specialists, especially from state and local control agencies where intrusion of federal authority was often unwelcome.[7]

A key industry ally in this debate was Abel Wolman (1892–1989), chief engineer of the Maryland Department of Health and chair of the Water Resources Committee of the National Resources Committee. Wolman's achievement in perfecting chlorination of public water supplies gave great authority to his advocacy of cooperation with industry in place of regulation. Before the Senate committee, Wolman made the failure of state regulation the starting point of his argument. Streams were no cleaner in states with strong control laws than in places without them; therefore little good would come of federal regulation. Wolman pointed as well to the expense of cleanup:

> There are many more situations, to my mind, where complete treatment is economically undesirable, is economically unfeasible. It may result in a somewhat less attractive stream, but the balancing of economic uses or advantages to the community may be greater than the salvaging of the stream for other less important purposes, and there, again, someone must make the decisions as to what those purposes might best be.[8]

Wolman left the health department in 1939 to begin a dual career, spending as much time as a consultant to a multitude of private and public organizations as on his faculty position at Johns Hopkins University.[9] He soon obtained a highly paid consultancy for Bethlehem Steel's plant just outside Baltimore. There he used his prestige and extensive political connections to lobby for Bethlehem, which for many years continued to dump untreated wastes into the air and water. In his old age, Wolman explained that he had

> agreed with [Bethlehem's] underlying assumption. It was not fair to ask Bethlehem to finance outlays in pollution control devices if the investment return did not compare favorably with the return on other capital projects. Net return was a fact of business life. "I have to remind people that industry is not a philanthropic institution. Some people mix them up."[10]

Not all scientists shared Wolman's views. A prominent chemist and toxicologist with left-wing sympathies, Max Trumper,[11] warned of the dangers of recently developed synthetic chemicals at an April 1936 conference on the Delaware River: "I do not say that cancer is caused by any one of the many chemicals in our polluted water but it is a difficult problem to solve what is the cumulative effect over a period of years..." Trumper saw old obstacles blocking action against these new problems:

Industry invariably protests because of the excessive costs...May I quote from a speech of Mr. W. W. Carr, Esq.: "The statute provisions are ample to protect the Schuylkill River from pollution; the difficulty lies in their enforcement, and they have remained almost entirely without life upon the statute books—an attempt was made to increase the penalties but the manufacturing interests kept it from passing..." The significant thing about this statement is that it was made in 1886 and referred to conditions of 1828 and 1832.[12]

———

The chemical industry reacted quickly to the threat of federal controls. Sheppard Powell, professor of sanitary engineering at Johns Hopkins where he was soon to welcome Abel Wolman as a colleague, spoke on behalf of the Manufacturing Chemists' Association at the March 24 hearing. Powell conceded that not enough had been done about water pollution—indeed, he confirmed one of the main contentions of the Lonergan bill's backers. Among the obstacles to progress he listed was "unfair competition resulting from enforcement of strict regulations concerning stream pollution on industries in certain States, while similar industries elsewhere were not so penalized." But "drastic regulatory laws" were not the solution. There was no need at all for new federal regulation; matters should be kept in the hands of the states. How the unfair competition problem would be solved, Powell did not try to explain.[13]

With Powell's testimony, the industry was just starting to go to work. Following the two-track approach of lobbying and voluntary self-regulation that had brought the API success a decade earlier, the MCA within months established a permanent Stream Pollution Committee. The committee started out by preparing testimony—modeled directly after the API's—for the next round of hearings. In this testimony, the pollution issue was framed as a question of efficient utilization of water resources. Waste disposal was a legitimate use entitled to share the waters with other uses such as water supply, power, transportation, and recreation.[14]

The next task was to prepare a resolution for the MCA executive committee that would set out an industry position on regulation. The committee prepared a very conservative document whose core principle was that "No regulatory program shall include a prohibition against the discharge of any waste...unless there is available a practical and reasonable method of treatment..." In other words, production came before pollution control. Regulators could not stop discharges but could only require that they be treated, and even that only when treatment methods were available at reasonable cost. In a very slight departure from the API strategy, states were to have sole oversight authority in nearly all situations but not invariably; a very limited scope of federal action would be granted in the case of interstate waterways that were not covered by interstate compacts.

Subordinating public health to production in this way was too much even for Lammot du Pont. A fierce opponent of government interference with business—he was then in the midst of the Liberty League campaign against Roosevelt's reelection—he also had a strong commitment to the safety of his company's plants. Du Pont had the resolution modified so that the restriction of state regulatory powers would extend only to existing plants. Control agencies would be empowered in the case of new plants to go beyond merely requiring whatever treatment might be affordable and insist that they make their wastes "innocuous."[15]

———

The fight over water pollution control went on through four years of legislative back-and-forth. Through many twists and turns, the central issue was clear: would the federal government have the power to require industry to abate discharges? Kenneth Reid, by now head of the Izaak Walton League, denounced the Barkley bill as "worse than useless" because it lacked mandatory controls. Industry insisted in public that it shared the objectives of the proponents of regulation, but its supporters conceded that the reality was otherwise. An Ohio official replied to Reid in the *New York Times* letters column:

> Drastic stream pollution legislation has failed year after year because it was largely guided by ardent, idealistic sportsmen imbued with more sentiment than common sense. It is perfectly natural that industry, almost to a man, has opposed every bill that mentioned pollution abatement, as a matter of self-defense.

Even the vocabulary of this state regulator—federal regulation is "drastic"— mirrored the API strategy memo.[16]

The Roosevelt administration at first leaned toward the voluntary approach, and the Barkley bill passed both House and Senate in 1936, but Lonergan managed to kill it with a parliamentary maneuver in the closing hours of the Congress. An industry-friendly bill passed the House again in 1937, while the Senate approved a bill that compromised between the Barkley and Lonergan approaches. Manufacturers flooded Congress with letters of protest against the Senate's action, and the conference committee deadlocked. Eventually the Senate caved in, and in June 1938 the two houses approved what was essentially the Barkley bill. Conservation groups demanded a veto; Roosevelt killed the bill with a pocket veto while avoiding a clear position on the main issue in dispute. The president justified his action by pointing to a clause that

Kenneth Reid, executive director of the Izaak Walton League and leading advocate of water pollution control in the 1930s and 1940s. (Courtesy of the Izaak Walton League of America.)

allowed the surgeon general to submit sewage treatment projects directly to Congress—this bypassed his newly established budget-making process.

The water pollution debate was renewed in 1940, and in that year the effort to enact a national system of environmental controls reached a high-water mark that would not be revisited until the 1970s. A revised version of the Barkley bill, with the objectionable budgeting clauses excised, first passed the Senate. In the House, an amendment adding federal regulatory authority over new pollution sources was offered. Roosevelt altered course to endorse regulation, and the amendment was adopted. But in conference committee the House and Senate each insisted on its own position, and the bill died.[17]

Halfway through the next Congress, Pearl Harbor put pollution control on the back burner along with other peacetime concerns. The war-driven expansion of industry, accompanied by the introduction of new chemical products, caused many problems to flare up, but they were generally dealt with at the local level or through the economic controls run by the Army, Navy, and War Production Board.[18]

———

At the conclusion of the Second World War, attention returned to the unfinished business of controlling water pollution. In the interim, the scale of the problem had increased and its nature had broadened. The vast industrial expansion of the war had sent new wastes into the nation's rivers, with expenditures on control constrained by the rushed schedules and material shortages of wartime. By 1948, the wastewater sent into the nation's streams consisted of 80% industrial waste and only 20% sewage.[19]

Clean water advocates quickly swung back into action. The Izaak Walton League, its influence growing with the renewed popularity of outdoor recreation, issued a call to action in early 1946 and lobbied for federal regulatory authority. Bills introduced in Congress reflected the same range of opinion as before the war. A compromise bill was approved by a House committee later in the year, but it did not reach the floor and died when Congress adjourned.[20]

Sensing the growing desire for cleaner water among the electorate, some within industry viewed more stringent regulation as inevitable and even desirable. Among them was Wilson Hart, who was in charge of waste disposal at Atlantic Refining and chaired the American Petroleum Institute's committee on refinery wastes. In January 1946, Hart began a series of technical articles in *National Petroleum News* with a call for a change of policy:

> ...many industries spent much thought, time, and money in a manner which only created the impression that they were trying to avoid doing anything at all about water pollution...This money could have been spent more profitably devising means for waste treatment.
> ...What has been said in considerable detail is intended to point out to industry, and particularly the refining industry, the futility of adhering further to the policy of objection and obstruction.[21]

Wolman and Powell also returned to this topic, speaking at an industrial waste symposium organized later in the year by Powell for the American Chemical Society. Wolman again made the case against federal regulation. Polluted streams, he said,

> are peculiarly problems for current diagnosis and treatment by state and local subdivision. The blanketing of the country by uniform law or by uniform administrative decision cannot lead to anything other than a wasteful abuse of a valuable material resource in our surface waters.

Powell, who by now had left Hopkins to consult full-time for industry, led off a long series of papers on water treatment technology. Saying that he would not comment on pending legislation, he nevertheless urged a middle course by implication. Criticism of "drastic measures" urged by control advocates was balanced with a warning to his industrial listeners that their wastes needed more attention and funding. "Management," he insisted, "must revalue the importance of recovery and treatment of wastes and give this phase of manufacturing a rank commensurate with production itself."[22]

The opinions of experts might be shifting toward control, but the political winds in Washington were blowing in the opposite direction. After pro-business Republicans took control of Congress in the 1946 election, industry stood firm in its opposition to regulation. Even the activist surgeon general, Thomas Parran, was forced to trim his sails. In an introductory talk at the Chemical Society symposium, Parran stepped back from the Administration's 1940 support of federal regulation. He stated the problem once again in stark terms:

> Waste products from our industrial processes constitute the largest source of pollution. While they differ in nature from domestic wastes, industrial wastes in some respects have a greater detrimental effect . . .
>
> Industry has not accepted fully its responsibility . . . The American people should not continue to tolerate the present gross pollution of our public waters.

But after unequivocally rejecting industry's usual excuse that "certain industrial wastes cannot be treated," he abstained from expressing any opinion about federal regulation and enforcement. Bowing to political realities, Parran said merely that it was a policy question to be decided by Congress.[23]

A year later, Oscar Ewing got rid of the independent-minded Parran, and soon thereafter the federal Water Pollution Control Act of 1948 was enacted. The 1948 law resolved the prewar struggle over the nature of federal involvement on the terms that had been sought by industry and industry's allies in state regulatory and public health agencies. Mirroring the Barkley bill of 1936, it explicitly reaffirmed state primacy in water pollution control and created no real federal enforcement authority. The central provisions authorized grants for sewage treatment plants—for which no money was ever appropriated, because the appropriations committee felt that the allowable sums were too small to make a difference—and research.[24]

In retrospect, the 1948 act is often described as a forerunner of today's complex regulatory system, the first federal law to address water pollution comprehensively. But it was, at best, an opportunity missed. The severe verdict pronounced by Kenneth Reid—"5 years hence we will be forced to admit that the present bill was a failure"[25]—turned out wrong only in the timing. Like the pesticide act enacted the previous year, the Water Pollution Control Act manifested the federal government's refusal to invoke its powers against the growing problem of chemical pollution.

CHAPTER 10

Deregulating California's Water

. . . a backlog of water pollution over the State that will constitute a plague comparable
to the air pollution in Los Angeles.

—Wilton Halverson, 1948[1]

WITH THE FAILURE OF the drive for federal regulation, control of water pollution remained in the hands of state and local agencies. Those who wished to clean up the streams had no choice but to lean on these weak reeds, and they sought what improvements they could get from state legislatures. Industry fought back, still seeking to fend off outside oversight, but a strategy of simple negativism was untenable—the glaring insufficiencies of existing control mechanisms had been exposed in the national debate. A burst of activity ensued in state capitals around the country.

Alongside the long-standing plague of stream pollution, the progress of the chemical industry brought new problems to underground waters. The knowledge that groundwater could be polluted was of course not new; as we have seen, court cases on groundwater contamination are attested in Europe as early as 1349. By the start of the twentieth century, public health authorities were enforcing the separation of wells from privies and other sources of pollution. Scientists of the 1920s and 1930s extended and systematized this knowledge.[2]

But the water pollution debates of the early decades of the twentieth century focused mostly on surface waters. Aquifers are less vulnerable than streams and lakes to the pollutants that were of greatest concern in those years. Sewage and food-processing wastes are readily digested by microbes in the soil; they pollute groundwater only when the water table is so shallow, the soil so sandy, or the flow so heavy that the waste is not fully consumed. With the new synthetic chemicals, nature often failed to offer this protection; indeed, many of these compounds were prized by industry for their resistance to microbial attack.

The breakneck industrial expansion of the Second World War, accompanied by the spread of synthetic chemicals, multiplied incidents of groundwater contamination. In New York, Michigan, California, and Maryland, chromium was found in water supply wells.[3] Gasoline leaked from underground storage tanks in Michigan and Virginia in such large amounts that scientific investigations were undertaken.[4]

The spectacular "Montebello incident" of 1945 demonstrated the vulnerability of groundwater to pollution by the new synthetics. A newly opened pesticide factory near Los Angeles dumped a bad batch of 2,4-D into the sewer. Raw materials used in making the weed-killer passed unaltered through a treatment plant that discharged its effluent into the Rio Hondo, a tributary of the Los Angeles River that drains the San Gabriel Valley. The effluent flowed several miles downstream through a rocky narrows and percolated into the ground when the river reached the sandy coastal plain. Foul-smelling contamination reached a public water supply well within three days, and after 17 days all of the city of Montebello's 11 wells were closed. The bad odor and taste, detectable at concentrations as low as a few parts per billion, persisted in the wells for four to five years.[5]

———

With water as with air, rapid industrial growth and unique environmental circumstances put Southern California in the forefront of the battle to bring pollution under control. Shortage of water, population growth, and industrial development combined to make any threat to water resources a public concern. As early as the 1920s, control of wastewater from oil wells and refineries had been a major political issue in the area, and sewage treatment works grew rapidly in number.[6] Southern California's dry climate made the protection of underground waters a matter of urgency. In a desert where rivers percolate into the earth rather than flowing to the sea, fish and game are sparse and streams have little value for industry or recreation. Groundwater aquifers are essential sources of water, and they are easily endangered by discharge of industrial waste.[7]

The wartime pollution of city water wells in the Los Angeles basin by chromium and pesticide wastes raised alarms among California health officials, and with the coming of peace they took action. The State Health Department began to implement California's old but little-used pollution control statutes. Laws passed between 1907 and 1917 required a permit to dispose of wastes of any kind—not just into streams, but also onto the ground in any location where the water percolated into underground drinking-water supplies. Permit holders were required to install treatment plants designed and managed to the satisfaction of the state, and the penalties for illegal dumping ran up to one year in prison for each day the disposal continued.[8]

In 1946, the Health Department issued new regulations aimed at enforcing these long-ignored laws.[9] The state's industries quickly mobilized against the threat of an end to cheap waste disposal. Their aim was to control the public debate over this complicated subject. As a first step, the legislature was convinced to appoint an investigative committee under the chairmanship of a pro-business Republican assemblyman, Randal Dickey. With weaknesses in the existing statute—most notably the absence of any sanction against violators short of criminal punishment—universally acknowledged, there was little disagreement that new legislation of some kind was needed.[10]

Business interests, the chemical manufacturers among them, set up a new organization called the California Association of Producing Industries to represent them in the water pollution debate.[11] The CAPI emulated the strategy that had been pioneered by the oil industry in the 1920s when it pinned the blame for pollution on bilgewater discharged from ships so as to preserve its own freedom to send wastes into the ocean. Working through the Dickey Commission, industry sought to shift the focus of attention away from industrial wastes and toward inadequacies in the treatment of household sewage.

In the spring of 1949, after 24 days of public hearings that attracted crowds throughout the state, the Dickey Commission issued its findings in a report that at first reading seems surprisingly contemporary. Its introduction observes that sewage discharges "caused a great apprehension on the part of domestic, agricultural, industrial, and recreational water users over the threat to the quality of their vital water supplies from unreasonable practices in waste disposal. The great wartime expansion of processing industries in the State caused even greater concern, particularly over the possible pollution of underground waters by mineral wastes." There was a widespread understanding that many industrial wastes are not broken down in the soil as sewage is, and that these materials are prone to migrate downward and pollute water supplies. Scientists, public health officials, and the public were intensely concerned about toxic chemicals getting into groundwater.[12]

But the Commission directed its focus away from these scientific findings. It instead emphasized two points that justified a permissive approach to industrial waste disposal. First, it stressed the economic importance of industry to California.[13] Second, it argued that direct threats to public health from water pollution almost always came from sewage rather than industrial waste.[14]

Notwithstanding its insistence on the generally innocuous character of industrial wastes, the Dickey report distinguishes with some care among the disposal problems of different industries and shows an awareness of the difficulties of handling chemical wastes. The wastes from the chemical, metal fabrication, and steel industries are said to be "of concern almost entirely as they may affect the domestic or agricultural use of underground waters." The assertion that few industrial wastes are hazardous to the public health is qualified with a mention of "the remote possibility of poisoning by some of the chemical wastes..."[15]

With this report as a basis, a thorough deregulation of industrial waste disposal was proposed. Dickey on January 27 introduced a series of bills whose fundamental premise was the principle enunciated by the Manufacturing Chemists' Association in 1936: Waste disposal is a legitimate beneficial use of water resources, to be encouraged in the interest of economic development. "Unnecessary costs and expense," the report argued, "are imposed upon some communities and industries by undue restriction on *reasonable use* of land and water for *necessary* waste disposal purposes. Orderly economic development in several regions of the State is being jeopardized for this reason."[16]

Dickey called for a repeal of the system of licensing by the Health Department, whose role was to be limited to emergency action in cases where people were provably sickened by polluted water.[17] Dischargers of waste were required merely to file reports describing their discharges with newly created Regional Water Pollution Control Boards. These five-member boards, appointed to represent industry, agriculture, water suppliers, and local governments, were largely autonomous; a state board was also established, but it had little more than advisory powers.[18]

Consistent with the views of the chemical industry, the state would not establish any general standards for water quality. Rather, dumping would be curtailed on a case-by-case basis when it was shown to interfere with previously existing uses of the water, such as drinking-water supply, irrigation, or recreation. As originally proposed, Dickey's legislation allowed the state to intervene only after finding an imminent hazard to public health; an adverse and unreasonable effect on domestic, industrial, agricultural, navigational, recreational, or other beneficial use; or odors or unsightliness resulting from unreasonable waste disposal practices. Under this standard, no action could be taken even in the case of a foul-smelling river unless the waste disposal that caused it was shown to be "unreasonable."[19]

Opposition to Dickey's proposals was led by the California League of Cities, upset by provisions overturning existing local government controls on industrial waste disposal, and by state and county health departments. They were joined by an alliance that included the Farm Bureau, water utilities, irrigation districts, and county governments. A competing bill, retaining licensing of industrial waste discharges by the State Health Department while creating new enforcement mechanisms, was drafted by Governor Earl Warren's administration and introduced by State Senator Nelson Dilworth.

After the Dickey bills were pushed through the state Assembly with little discussion, matters came to a head on June 7. Governor Warren denounced the bills in a press conference, describing them as an industry-backed proposal that would put the polluters in charge of the pollution control setup. At the same time, Health Department director Wilton Halverson testified before a committee hearing that attracted such a large crowd it had to be moved to the Senate chamber. Halverson said the plan would create "a backlog of water pollution over the State that will constitute a plague comparable to the air pollution in Los Angeles." He urged the Senate to retain state permitting, issuing a prescient warning that prevention was the key to controlling water pollution.

Dickey struck back with an attack on his critics' motives, accusing local governments of causing 85% of the problem by inadequately treating sewage. The cities, he claimed, favored the Dilworth bill's permit system only because it would grant them a license to continue polluting. This was a transparent falsehood; the League of Cities had recently won state funding of sewage treatment over opposition that included Warren himself.[20]

Charge and countercharge flew for another week, until a compromise was brokered by the Los Angeles Department of Water and Power. The Dickey proposals had been changed earlier to allow local governments to license waste discharges if they chose, a power already exercised to good effect in Los Angeles and Orange Counties. The new regional water boards were now authorized to impose requirements on the contents of wastewater before it was released and diluted in the environment.[21] But these limited regulatory powers were severely circumscribed. Discharge requirements had to prescribe limits on the contents of the outflow, without specifying treatment methods. This meant that proving a waste to be toxic did not enable a board to restrict its discharge; the specific substances causing the toxicity had to be detectable in the waste by the analytical chemistry of the day. Requirements could be appealed to a statewide board. And even when requirements were imposed, they were not easy to enforce. The board bore the burden of proving in court that existing uses of the water had been curtailed as a direct result of the violation.[22] The amended bills passed quickly and were signed by Governor Warren a month later.

———

The new system of pollution control took shape over several years. In Los Angeles and Orange Counties, the previous system of regulation remained intact; the water board reviewed local industrial waste permits before issuance and then issued as its own "requirements" a simple statement that the discharger must comply with its municipal permit.[23] Elsewhere, the boards were in need of guidance for making the case-by-case determinations that the new law imposed.

The State Water Pollution Control Board took this issue up at its first meeting in December 1949 and appointed a Committee on Water Quality Criteria. The committee reported after three months with a table listing acceptable levels of contaminants for various water uses. This chart elicited sharp criticism. The board, reaffirming that California had chosen not to impose water-quality standards, rejected the table, and instead commissioned a compilation of data on a wide spectrum of contaminants. This job was assigned to Jack McKee (1914–79), a specialist in water treatment technology. McKee was a founder in 1947 of the water treatment consulting firm Camp Dresser McKee. In 1949 he joined the faculty of Caltech, developing the school's new program in sanitary engineering while continuing to consult for his firm's municipal and industrial clients.[24]

McKee's compendium was published by the state board in 1952. Entitled *Water Quality Criteria*, it was deeply infused with the policy that the chemical industry had advocated since the '30s and California had enacted in the Dickey Act—to promote waste disposal as a beneficial use of public waters. The deterioration of water quality, McKee later wrote, was a trend that "cannot be stopped or reversed...unless the industrial and agricultural development of this Nation is to be curtailed."[25] *Water Quality Criteria*, and through it the philosophy of the Dickey Act, came to influence water pollution control practice far beyond California's borders. Yet this publication owed its official imprimatur and the influence that flowed from it not to its scientific merits but to a political decision of the California legislature.

In assembling his scientific information, McKee followed the "innocent until proven guilty" approach that Robert Kehoe had introduced in his studies of tetraethyl lead and that industry-friendly scientists had subsequently applied to synthetic pesticides. Only those studies that determined "limiting or threshold concentrations" were included in the compilation. When no minimum concentration was known below which a substance did not cause harm—as in the case of cancer-causing chemicals—McKee often omitted the existence of the hazard altogether and declined to offer warning of a danger from which no safe harbor was known. Beyond that, the dangers of industrial chemicals were often downplayed. Thus the entry for chromium failed altogether to mention the high rates of lung cancer among chromium plant workers, which by 1952 had been clearly established. And, after stating the drinking-water standard for chromium that had been set by the Public Health Service, McKee commented, "Recent observations tend to discount the fears of the U.S.P.H.S. and the foregoing statements relative to the physiological effects of hexavalent chromium."[26]

As McKee pointed out in the transmittal letter of his compendium, he avoided using the term "standards" to show that no definite rules were being set. He explained this in a talk at a scientific conference several years later:

> The fact that it has been established by authority makes a standard somewhat rigid, official, or quasi-legal; but this fact does not necessarily mean that the standard is fair, equitable, or based on sound scientific knowledge, for it may have been established somewhat arbitrarily on the basis of inadequate technical data tempered by a cautious factor of safety....
>
> In contrast, a "criterion" designates a means by which anything is tried in forming a correct judgement respecting it. Unlike a standard it carries no connotation of authority other than that of fairness and equity.[27]

To McKee a pollution control expense occasioned by unneeded safety factors would not be fair. No such unfairness arose when the public was denied knowledge that cancer-causing chemicals were in the water they drank. In placing the financial interests of waste dischargers ahead of the health of those exposed to the wastes, *Water Quality Criteria* encapsulates the spirit of the Dickey Act.

The next major California water pollution law, the Porter-Cologne Act of 1969, passed an implicit judgment on the Dickey Act by repealing the most controversial provisions passed 20 years before. Waste could be discharged only after advance notification of the Water Board and issuance of requirements, in effect returning the state to a permit system.[28] The concept of waste disposal as a beneficial use of waters was explicitly rejected.[29] Board orders could be enforced without proof that an existing use of water would be affected.[30] That the Porter-Cologne Act, which remains to this day the foundation of water pollution control in California, could be passed by unanimous vote of the legislature with the support of Governor Ronald Reagan is the measure of the backward step taken 20 years earlier in 1949.

New York was another state where the Second World War brought pollution to both surface water and groundwater. The rapidly expanding suburbs of

Long Island, built on highly permeable glacial sands where almost all rainfall percolated into the earth, relied on wells for their water supply. As in California, an aircraft manufacturing industry arose on the hard, flat terrain where airports were easy to build.

And like California, New York had its spectacular groundwater contamination. During the war, wells near two aircraft plants on Long Island were closed when they were polluted with chromium. The two factories discovered their problems in very different ways. At Liberty Aircraft, county health officials who observed chromium wastes poured into a pit took the precaution of testing a nearby private well. The well was closed when its water was found to contain 100 parts per billion of chromium. But the second case, one year later, was detected only when workers at a Grumman Aircraft plant in Bethpage noticed yellow water in their drinking fountains.

New York City, which drew some of its water supply from a point three and a half miles to the south, installed seven shallow test wells near Liberty Aircraft in 1945 and initially found no contamination. When the war ended, further investigations began, with dozens of new monitoring wells installed. In 1948, chromium was detected in three of the city test wells near Liberty, and a plume of contamination was traced for a mile from the plant. At Grumman, chromium had been found in public supply wells more than 100 feet deep. By 1949, all of Long Island's aircraft plants had installed treatment systems to remove chromium from their wastewater.[31]

Again like California, New York responded to these problems with a legislative inquiry that included a scientific survey of the problem and public hearings throughout the state. The task was more complex than in California, since many New York waterways flow into other states—New York is drained by the Hudson, the St. Lawrence, the Delaware, the Susquehanna, and even by tributaries of the Mississippi. The job was therefore assigned to the Joint Legislative Committee on Interstate Cooperation. Work began in 1946 under the leadership of Assemblyman Harold Ostertag, a Republican from rural Wyoming County.

Ostertag's committee painted the picture of water pollution in shades quite different from those Dickey had used. Industrial wastes received equal billing with sewage. While the committee repeated the usual platitudes about industry's desire to reduce pollution, the facts it compiled made clear that real progress was slow. Numerous sites of industrial pollution were mapped, with chemical plants in Buffalo seen as a particular problem. Aside from the Long Island chromium removal systems, only three major industrial treatment facilities were built anywhere in the state in 1948. The total construction cost of needed industrial waste treatment was put at $100 million in 1947, and the estimate quadrupled within three years.[32]

New York's existing system of regulation had grave defects. The Ostertag committee's initial progress report, issued in February 1947, summarized them. The Health Department had authority to halt pollution, but only after a considerable number of people had actually become ill, and industrial wastes were exempted unless mixed with sewage. A separate statute dealt with industrial waste—the Conservation Law, last amended in 1913, which specified that

> No dyestuffs, coal tar, refuse from a gas house, cheese factory, cream-ery, condensary or canning factory, sawdust shavings, tan bark, lime or other deleterious or poisonous substances shall be thrown or allowed to run into any waters, either private or public, in quantities injurious to fish life . . .

Notwithstanding the early date of the statute, its scope was broad and the chemical industry unquestionably fell within its ambit. Enforcement was difficult, however. A penalty of $500 plus $10 per dead fish could be assessed, but it was necessary to identify the substances that had killed the fish, show who put them in the stream, and prove that the substances were harmful to fish at the concentrations in the stream. To make these demonstrations, the state Conservation Department had hired chemists and biologists, but often the necessary proof could not be obtained. Even when it was, for large facto-ries the penalty was often no more than an added cost of doing business and the discharge would continue.[33]

Ostertag sought a consensus solution. The leaders of the Izaak Walton League and the state Conservation Council—constituencies conspicuously absent from the California debate—sat on his committee as advisory mem-bers alongside representatives of industry and local government. In Septem-ber 1947, the committee issued a draft bill for public discussion. The bill, which won the active support of conservationists, was introduced in the 1948 session of the legislature, but not voted on. The legislative recess was used by the committee to make revisions. After dealing with last-minute industry objections, Ostertag introduced the final version of his bill on February 28, 1949. It won the legislature's unanimous approval in less than a month and was soon signed into law.[34]

The 1949 law strengthened both the structure and the substance of New York's water pollution control program. A Water Pollution Control Board composed of cabinet-level state officials was established with broad regula-tory powers. Following a method pioneered by Pennsylvania in the 1920s, the board was to classify the state's waters according to their designated uses, from AA for pure drinking water to F for water bodies used exclusively for waste disposal. There were separate categories for tidal salt waters and groundwater. Enforceable quality standards were established for each category; the Board specified numerical standards for cyanide, ammonia, copper, zinc, and cadmium and added a blanket prohibition of "toxic wastes, oil, deleterious substances" in amounts that made waters unsuitable for the designated uses of each class. New waste discharges required permits, and the Board had authority to order new controls on existing discharges.[35]

———

The California and New York debates over water pollution were not isolated events. In New Jersey's Raritan River valley, a center of chemical manufacturing, two thousand people joined a Restore the Raritan Society in 1949. The state's Board of Health, which since at least 1909 had been pressing municipalities and factories along the river, placed increased pressure on chemical plants to treat their wastes, and disputes began to reach the courts.[36]

Legislative activity at the state level was widespread in this period. Some states followed New York's approach of setting water-quality standards and issuing permits; others adopted the method of case-by-case analysis that chemical manufacturers favored. Pennsylvania strengthened its controls in 1945 and then retreated under industry pressure two years later. Ohio emulated California; in 1951 the state created a Water Pollution Control Board whose seemingly vast powers were so encumbered by procedural fetters and technical hurdles that it was rendered impotent in practice.[37]

In addition, as historian Craig Colten has pointed out, during the late 1940s there was "a growing national discussion on the need to offer groundwater equal protection" with surface water. In Illinois, for example, a 1948 legislative report recognized the need to prevent groundwater pollution and identified a nationwide trend toward unified regulation of surface water and groundwater. By 1951, state law was amended to explicitly direct a reconstituted Sanitary Water Board to "control, prevent, and abate pollution" of groundwater from all sources, including industrial waste. Alabama in 1947 established a Water Improvement Advisory Commission for "the improvement and conservation of the ground and surface waters" and recognized industrial wastes as a major cause of pollution. In 1953, the word "advisory" was removed from the commission's name and a permit system for waste discharges was established.[38]

Regardless of the variation among states, essential elements of control remained missing in the 1940s. Uniform standards for effluent water quality were lacking, and enforcement of the standards that did exist was weak. Even in states like Pennsylvania and New York, a factory had to clean up its wastes only so far as the designated purpose of a waterway required. Manufacturers were left free to fend off treatment requirements by threatening to relocate. When standards were not met, persuasion was preferred to coercion. Water treatment practice would steadily improve, but at a pace that failed to keep pace with the growth of industry and the emergence of new contaminants.[39]

CHAPTER 11 | The Stealth Pollutants

a sampling might immediately show pollution from past activities ... discretion is the better part of valor

—Robert Brenneman, Lockheed Aircraft, c. 1979[1]

SYNTHETIC ORGANIC CHEMICALS OF many kinds came into wide use during and after the Second World War. Among them, pesticides were far from alone in doing environmental damage. Another class of materials that turned out to have a wide impact was the chlorinated solvents. Specialists were aware very early of the toxicity of these chemicals, but their propensity to pollute was slow to draw attention. They are volatile chemicals that pass quickly from streams into the air and rarely contaminate surface water supplies. Only when poured into the ground do they persist, leaving underground water supplies poisoned for decades.

Manufacturers of these compounds warned their customers to avoid practices likely to spread them through the earth. But, fearing to draw attention, they offered little explanation of the reason for the warnings. The invisibility of the danger gave industry the luxury of inaction. The pollution hazards of chlorinated solvents gained public notoriety only when discovered in drinking-water wells throughout the country in the 1970s. Today, chlorinated solvents are widely thought to be the most troubling class of groundwater contaminants. The history of their spread is a case study in the weakness of self-regulation.

———

The term chlorinated solvents describes compounds that consist of a short chain of carbon atoms with chlorine attached. Carbon tetrachloride and trichloroethylene (TCE) are perhaps the best-known members of this group. The labeling of these compounds as solvents is something of a misnomer; they have many uses. They are indeed solvents—used for degreasing, dry cleaning, and many other purposes—but they have a multitude of other applications, including pesticides, chemical intermediates, and refrigerants. The greatest volume of use has been to remove grease from metal parts in aircraft manufacturing and other metal-fabricating industries.

Like many synthetic compounds, the chlorinated solvents were first produced commercially in Germany. In the United States, manufacture began on a small scale before the First World War, taken up at sites where chlorine, then in surplus because it was a by-product of other manufacturing processes, was available. Dow Chemical, first established to extract bromine from brines, had begun early on to extract chlorine as well. Dow decided even before the First World War to invest in research on chlorine compounds, and it became the first major American manufacturer of chlorinated solvents. In the 1920s and 1930s it was joined by DuPont and several other producers.

First of these compounds to come into wide use was carbon tetrachloride. Highly effective in grease removal, dry cleaning, and similar applications, its lack of flammability offered a great advantage over gasoline, ether, and other competing solvents. Sales rose rapidly during the 1920s and 1930s. Another member of the class, chloromethane, found early use as a refrigerant.[2]

The dangers that these compounds posed to health soon became evident. The Chicago refrigerator deaths were the best publicized incident, but far from unique. In the early 1920s, cattle in Europe were killed by trichloroethylene used to remove fat from soybean meal, and evidence for the toxicity of TCE accumulated in the 1930s. The dangers of chlorinated solvents were well enough known among environmental scientists that the Los Angeles Air Pollution Control District, when presented in 1948 with the riddle of smog-caused damage to crops, initially suspected chlorinated solvents as the damaging agent. Arie Haagen-Smit looked for them and did not find them; when he instead detected peroxides, which could only be produced by oxidating petroleum constituents, the way was opened for the discovery of photochemical smog.[3]

In 1941, scientists at the National Cancer Institute discovered that carbon tetrachloride, when fed to mice, causes liver cancer. The need to follow up by testing similar compounds was obvious, but work proceeded slowly—the institute was so shorthanded that it could not test DDT when asked by the Army Surgeon General. By 1945, the experiment had been repeated with chloroform, yielding tumors that were indistinguishable from those produced by carbon tetrachloride.[4] The implication was clear: chlorinated aliphatic hydrocarbons—the class of solvents that includes carbon tet and chloroform—should as a group be suspected of causing cancer. Wilhelm Hueper did not fail to point this out in the publications he issued after joining the NCI in 1948.[5]

Contemporaneously with these discoveries, the uses of carbon tet that led to the greatest human exposure were phased out. The main substitutes were TCE for degreasing and perchloroethylene (PCE) for dry cleaning. The chemical literature attributes this shift, in part, to toxicity; because the manufacturers' internal files have not been made public, it is not clear whether the cancer findings figured in the decision.[6] The two chemicals that replaced carbon tet are also chlorinated aliphatics, but (as far as is known) none of the manufacturers chose at the time to test whether TCE and PCE are carcinogens. The means for such testing were at hand; by the time TCE and PCE came into heavy use, Dow and DuPont had operated their own toxicity laboratories for more than a decade. A summary of the DuPont laboratory's standard services distributed to the operating divisions in 1951 offers tests for carcinogenicity and states that they "must, of course, be carried out whenever there is a clear indication for their use…"[7] Not until two decades later, in the early 1970s, would such testing be carried out—showing, as Hueper had feared, that both substances cause cancer in laboratory animals.

———

The first recorded detection of chlorinated solvents in drinking-water wells was in 1949. A pair of British public-health chemists, Francis Lyne and Thomas McLachlan, described two instances of TCE contaminating groundwater. Their brief report was published in *The Analyst*, one of the world's leading journals in analytical chemistry. In one case they described, waste TCE was dumped 150 yards from a well; the concentration in the well was measured at 18 parts per million. Lyne and McLachlan commented that "it is evident that contamination by compounds of this nature is likely to be very persistent and there is some evidence of toxicity at very low concentrations."[8] The report by Lyne and McLachlan attracted some notice when it appeared and continued to be cited in subsequent years, but it was not highlighted as a groundbreaking discovery.[9] The news that TCE contaminates groundwater when put into the ground was, it would seem, received by scientists as an unsurprising confirmation of what was naturally to be expected.

The toxicity, chemical stability, and heavy use of chlorinated solvents signaled to attentive scientists and engineers that these compounds would pose a more widespread contamination problem. California water official Harvey Banks warned in 1952 that:

> Serious conditions of pollution of the ground-water resources occur when industrial wastes containing organic solvents and other chemical compounds are either indiscriminately discharged into open unlined sumps or directly onto a highly absorptive surface dump site area...Solvents and soluble chemicals contained in industrial wastes...remain in solution in the liquid phase in most instances and percolate downward to the underlying ground water.[10]

Banks was not the only water pollution specialist who worried about these chemicals. One of California's newly established Regional Water Pollution Control Boards in 1952 forbade a Sacramento rocket-fuel plant to discharge wastes containing TCE or PCE "in a manner which will permit their entry into either the ground water or the waters of the American River."[11] In early 1953, the board asked the state Department of Water Resources to help monitor "any effect of chemical wastes from the plant upon ground water in the area." The investigation included TCE and PCE.[12]

The chemical manufacturers themselves perceived even more clearly than outsiders the dangers that chlorinated solvents posed to groundwater. Chlorinated aliphatic hydrocarbon compounds, some of them identical to materials used as solvents and all having properties similar to the solvents, were sold to farmers for use in the soil of their fields. The industry worked hard to understand how these chemicals behave in the ground.

The use of fumigants began after the First World War. When combat ended, there remained a large surplus of the chemical warfare agent chloropicrin, a tear gas that generally did not kill. Experiments showed that chloropicrin kills nematodes, tiny worms that infect plant roots. The physical properties that armies had exploited to deliver the compound on the battlefield were convenient for use in agriculture. The chemical could be stored in liquid form and injected as a liquid into the ground, where it would gradually evaporate and spread through the surrounding soil as a gas. Borrowing a term first used to describe disinfection in sealed chambers, this form of insecticide delivery became known as fumigation.

Fumigation turned out to be especially useful in growing pineapples, and it came into wide use in Hawaii by the mid-1930s. Chloropicrin, until the war surplus was used up, was the agent of choice; meanwhile a search went on for replacements. Research continued during the Second World War in several venues; chemical manufacturers, the Agriculture Department, and the grower-financed Pineapple Research Institute all contributed. At war's end, two new fumigants were available: D-D, a mixture of 1,3-dichloropropene and 1,2-dichloropropane, and ethylene dibromide. With the chloropicrin stockpile disappearing, they were quickly accepted by the pineapple growers.

The manufacturers sought wider markets; in 1946, a massive effort to market fumigants for use on other crops was launched. Potential customers had to be convinced they had a problem before they would pay for a solution. Nematodes are tiny animals visible only under a microscope, and farmers saw no need for expensive remedies against damage they had always lived with, coming from a pest they had never seen. Chemical companies, enlisting the help of the Agriculture Department's extension agents, undertook a program of demonstrations that lasted years. Eventually the farmers of an area, seeing the increased yields, would make the fumigant part of their routine.[13]

Developing and marketing fumigants required much research. One thing the manufacturers particularly wanted to understand was how quickly these chemicals spread through the ground. Experiments on this subject were under way by the 1940s and studies continued vigorously thereafter; the results were widely diffused within the pesticide industry. When fumigants are injected into dry soil 12 to 18 inches beneath the surface, it was learned, they evaporate readily and tend to stay in the shallow root zone rather than migrating deeper into the ground. When the soil is wet, however, evaporation is slower and downward migration more frequent.[14]

By the 1950s, the science was sufficiently developed for the manufacturers to understand that fumigant constituents could persist in the subsurface environment and travel with the water phase, percolating down through the soil toward the water table. A 1961 fumigant advertisement would boast that "Nemagon goes where the water goes—kills nematodes at great depths," and soon thereafter researchers would be measuring the depth of fumigant penetration.[15] Clearly, a chemical with these properties could reach the water table, and within a few years the manufacturers began to detect groundwater contamination. D-D was detected in 1964 in a spring and a well near a French vineyard; in 1968 it was found in a California well.[16] In the 1980s, extensive fumigant contamination was detected in public water supplies in Hawaii and elsewhere.

If fumigants could contaminate groundwater, there was a clear danger when similar chemicals were used as solvents. In agriculture, 5 to 50 gallons were typically applied per acre.[17] Dumping of spent solvents or leakage from degreasers pours chemicals onto the ground at far greater rates—a spill of 20 gallons in an area of 100 square feet is equivalent to an application rate of nearly 9,000 gallons per acre.

Solvents spilled in large amounts can flow downward through the soil without the need of infiltrating water to carry them. After reaching the water table, they may continue moving down; their low solubility hinders mixing with the water, and greater density makes them sink. Manufacturers seem to have thought about this even before instances of solvent-contaminated wells began to find their way into the scientific literature. In the Dickey Commission report, issued a half year before the September 1949 date of the Lyne and McLachlan paper, a table provided by the chemical industry describes the wastes produced by 23 different industries. Only two industries are named whose disposal plans must, under average conditions, consider that their waste "affects usefulness of water to fish, animals, or humans even after treatment." These are manufacturing of the weed-killer 2,4-D—no doubt included because it was the cause of the Montebello incident—and metal fabrication. Metal fabrication was also the only industry, other than oil and gas production and refining, in whose wastes the presence of an "immiscible component" was normal. This immiscible behavior—the chlorinated solvents act as what are now called "dense non-aqueous phase liquids"—causes chlorinated solvents to move deeper into the ground and linger there, making the pollution from these compounds more extensive and longer-lasting.[18]

———

Armed with the knowledge gained through fumigant research, the chemical industry sought to protect itself by warning the customers of its solvent businesses against unsafe disposal practices. An important channel through which these warnings were issued was the publication of Chemical Safety Data Sheets (CSDSs) by the Manufacturing Chemists' Association. The trade association began publishing these documents in 1947, and chlorinated solvents were among the first chemicals written up. Other compilers of safety information relied heavily on the CSDSs and often echoed their recommendations.

When it wrote the CSDSs, whose text was edited by the association's legal committee before publication,[19] the industry walked a fine line. These documents had a dual purpose: to inform and protect the users of chemical products, but also to protect the manufacturers from liability. Their language thus combines specific recommendations for disposal with cloudy explanations of the circumstances that make the recommended practices necessary. Crucially, the guidance insists that the solvent user fully consider all relevant site-specific circumstances prior to making disposal decisions, and seek expert advice when necessary.

The CSDS for trichloroethylene was among the first to be published. This 1947 document devotes just one paragraph to waste disposal, with a single sentence that addresses disposal on the ground: "Residue may be poured on dry sand, earth, or ashes at a safe distance from occupied areas and allowed to evaporate into the atmosphere."[20]

On a careful reading, these words raise as many questions as they answer. Why are dry materials specified? Why place any limits on disposal to the ground? Because the recommendation is placement on the soil to promote evaporation into air as the ultimate sink, these cautions cannot be intended to protect the soil or the air. What then is in the ground that must be protected? The only resource left that might cause concern is groundwater. The recommendation is for disposal by evaporation, and the chemical manufacturers knew from their fumigant research that dry conditions promote speedy evaporation. In wet ground, solvents evaporate more slowly, remain longer in the soil, and are more likely to migrate downward.

The TCE document was revised one year later, adding additional warnings and advice:

> Waste disposal of trichloroethylene depends to a great extent upon local conditions. Be sure that all Federal, State and local regulations regarding health and pollution are followed. The supplier will be able to furnish good advice.

The new language verifies that the advice on ground disposal arises from environmental and regulatory concerns. It also tells users of TCE to seek advice from the manufacturers when making their disposal decisions—there were other ways to get rid of wastes, including off-site disposal, recycling, and incineration. The waste disposal section remained essentially unchanged when the TCE datasheet was revised in 1956. Similar language on local conditions, regulations, advice from suppliers, and disposal onto dry soil is found in CSDSs for other chlorinated solvents, and for some nonchlorinated ones as well. In 1961, advice somewhat more explicitly aimed at protecting groundwater began to appear in CSDSs for some solvents.

Only in the 1970s did the vast extent of groundwater pollution by chlorinated solvents become apparent to regulators and the public. Industry was soon required—often with a reluctance exemplified by the quotation that begins this chapter—to find out what was beneath their properties. By the early 1980s, the widespread and persistent character of chlorinated solvent contamination was widely recognized. The enormous expense of cleaning up this mess has inspired much dispute about the causes. An entire literature has grown up to argue that the problem came as a surprise and could not have been foreseen. This contention relies heavily on the advice given in CSDSs, the reasoning being that disposal on the ground would not have been recommended if its consequences had been understood.[21]

But the CSDSs do not show any lack of knowledge on the part of their authors. Failure to warn does not prove ignorance, and an artfully worded warning proves it even less. The chemical industry's advice was not that risks of groundwater contamination were nonexistent, but that such risks could and should be kept under control by proper management. One of the industry's fundamental contentions, printed on the cover of every CSDS after 1948, was that "chemicals in any form can be safely stored, handled, or used if the physical, chemical and hazardous properties are fully understood and the necessary precautions...are observed." Even though chlorinated solvents pose inherent dangers, the manufacturers were saying, they can be safely placed on the ground under conditions chosen to promote evaporation and discourage downward seepage.

The carefully worded advice in the CSDSs is imperfect by modern standards, but it does reflect an understanding of the need to protect water resources. There is, indeed, no evidence that any significant portion of the solvents we find in groundwater today was put there by people who carefully followed these recommendations.[22] But when they offered their recommendations, the manufacturers disclosed just enough to protect themselves against liability. They failed to offer the specifics that could have motivated readers to follow their advice more closely. Once again, environmental protection fell by the wayside as the chemical industry pursued other priorities.

PART III | # Holding Back the Deluge

CHAPTER 12

DuPont Tries to Clean Up

At Baltimore a study is being financed by the Pigments Department to determine
the effect of waste on the water of Baltimore Harbor. This, in part, is an attempt
to develop factual information and is in part a delaying action to avoid requesting
the appropriation of money...

—Lyman Cox, 1950[1]

DUPONT WAS THE CHEMICAL industry's leader. The oldest, largest, and in many ways most technically advanced of American chemical companies, its prominence was magnified by the outsized role it took on the national political scene in the '20s and '30s. The federal government, in a 1948 antitrust suit, contended that DuPont's ownership of General Motors constituted the "largest single concentration of power in the United States." Company president Lammot du Pont served as president of the Manufacturing Chemists' Association from 1929 to 1933 and from 1939 to 1942, and then as chairman of its Board of Directors from 1942 to 1946.[2] DuPont is thus an archetype of industry's approach to pollution control.[3]

It was in the 1930s, as a result of the debate over the Lonergan bill, that liquid wastes and stream pollution became a matter of concern to DuPont's top executives. Following their laissez-faire philosophy, DuPont's executives had a strongly negative response to Lonergan's legislation, and they did not fail to make their views known to Congress. But while political action came first, it was an incomplete response to what was recognized as a real problem; what would the company do about its own plants? Lammot du Pont knew the issue would not disappear. A strong believer in fundamental research, he hoped for a technical solution: "It behooves the foresighted manufacturer, therefore, to anticipate this trend and to be ready for drastic legislation when, or even before, it comes."

DuPont in 1921 had adopted a decentralized structure in which general managers of manufacturing divisions held responsibility for the operation of their separate product lines. Under this structure, top management had limited power to impose its will. Its role was to set policies, and in 1938 DuPont's executive committee adopted a policy on "trade wastes":

> The committee considers this subject one of major importance and one which should receive continuous study of the same type as is applied to safety work and fire protection.

This was not a weak statement—fire protection was a pressing concern in what was still, to a substantial degree, an explosives company. But it was only a statement. In the hope of influencing the operating divisions toward responsible action, DuPont established a trade-waste staff that was housed in the corporate Engineering Department.[4]

These specialists' assignment was to provide technical support and oversight to the divisions. Technical help might be welcome, but oversight was not easy to implement. Lammot du Pont asked his staff to develop quantitative measurements of the divisions' environmental performance that he could use as a measuring stick. His model, surely, was the centralized financial accounting that Lammot's illustrious predecessor, Pierre du Pont, had developed to manage the company's rapid growth before and after the First World War. The comprehensive environmental measuring stick was an elusive goal, as policymakers were to learn repeatedly in later decades. The staff found it hard to exert direct influence on the divisions; its role in the 1930s turned out to be mostly technical support and information exchange.[5]

———

With the coming of war, DuPont's engineers and chemists turned their attention to the country's defense needs. Their work reflected the diversity of the company's businesses—extensions of existing product lines like nylon, rayon, insecticides, and of course explosives, and newer materials such as polyethylene and Teflon. These efforts, important though they were, are overshadowed by DuPont's role in building the first atomic bomb as part of the Manhattan Project. In December 1942, the company took on the assignment of designing, building, and operating the world's first plutonium production plant at

Hanford, Washington. The key task of designing this one-of-a-kind plant fell to the Engineering Department.[6]

The Manhattan Project marked scientists and engineers who participated in it for the remainder of their working lives. The designers and operators of the Hanford plant confronted threats to air and water of an unprecedented nature and scale. The job of containing those threats fell to DuPont's environmental specialists. Like others in the bomb project, they would not be the same when they emerged from the experience.

One obstacle was the highly radioactive gases released from plutonium separation. This problem was known from the earliest experiments with the process and was a major reason for the selection of the remote Hanford site. The plant was built with 200-foot stacks, and before it opened, a meteorological team studied the site's changing wind patterns in great detail. The team made 36,000 individual wind measurements and ran dispersion tests with oil fogs. Radiation detectors were installed in nearby towns. To avoid excessive releases of radioactive iodine and xenon, the meteorologists determined, fuel discharged from Hanford's reactor would have to be held at least 30 days before it could be dissolved in the separations plant. The purpose of the delay was for radioactive decay to destroy iodine-131, a dangerous thyroid poison; this isotope's eight-day half-life means that 90% will be gone after a month's wait. To further protect the surrounding population, the fuel was to be dissolved only when wind conditions favored dispersion into the atmosphere. Daily weather forecasts were phoned in to plant managers who planned the dissolution campaigns.

Under extreme pressure to produce the first bombs, these precautions were not always followed. In the first half of 1945, the holding time was sometimes less than a month, and in the second half of the year dissolution was carried out under unfavorable wind conditions 40% of the time. Iodine-131 was detected off-site in the spring of 1945, and in the fall a study of its accumulation on vegetation was begun. In June 1945, routine thyroid checks of site employees were begun. In December, with the war won, the holding time was increased to 60 days, a delay sufficient to eliminate 99% of the radioactive iodine.

The challenges posed by the uniquely high concentrations of radioactivity in the plant's liquid wastes were, if less immediate, even more daunting. These wastes emerged from the plant as highly corrosive nitric acid solutions, exacerbating the difficulty of handling. The three most dangerous waste streams, it was determined, could not be disposed of like other chemical wastes, but must be kept in tanks permanently, or at least until a better solution could be devised. Stainless steel, which could resist corrosion by

The first high-level waste tanks under construction in 1944 at the Manhattan Project's plutonium plant in Hanford, Washington. (Courtesy of the U.S. Department of Energy, Hanford Declassified Document Retrieval System.)

the acidic wastes, was in extremely short supply during the war, so carbon steel tanks were constructed and the high-level wastes were neutralized. This treatment caused most of the radioactivity to clot up into sludge and settle at the bottom of the tanks.[7]

DuPont, which had been stung by New Deal–era investigations into the large profits the company had made during the First World War, decided not to make atomic energy part of its business. The company did its work at Hanford for a fee of one dollar and renounced all patent rights. (This principle of renunciation did not apply when future nonmilitary markets were at stake. According to the army surgeon general, DuPont, which had lagged in pesticide research before the war, refused in 1943 to manufacture the new insecticide DDT for the military until the patent holder, Geigy, agreed to license it for postwar civilian production.) When hostilities ended, the company gave up its contract at Hanford, and in September 1946 the Engineering Division returned to civilian work.[8]

Hanford's war-era high-level wastes remain to this day in the same steel tanks. With the tanks now leaking and the bulk of their radioactivity contained in hard-to-remove sludges, they constitute a massive environmental problem that remains unsolved. The wartime exposures of Hanford-area residents to radioactive iodine were also far greater than was safe. The decisions that created these conditions have been much criticized in retrospect. But when viewed in the context of the time, there is much to praise in DuPont's environmental record at Hanford. Problems were anticipated before they arose, and intensive efforts were directed to their solution. The most dangerous wastes were put into tanks—a measure that later proved to be inadequate, but one that went beyond the usual industrial practice of the day. After DuPont's departure, the site continued to use single-shelled tanks until 1964 despite a lesser urgency of production and the renewed availability of stainless steel. In a few cases, wastes of types that had been stored in tanks during DuPont's tenure were later poured onto the ground.[9] DuPont, in contrast, learned from its mistakes when it returned to the atomic energy field in 1950; high-level wastes were identified as a major issue at the very start when the company planned a new plutonium plant on the Savannah River in South Carolina. The South Carolina wastes were put into double-shelled tanks; by 2006, many of the older tanks had been cleaned out and only one had leaked into the environment.[10] Environmental decision making at Hanford was a shared effort with Manhattan Project scientists, but the chemical company was in control of the plant design; without the commitment to environmental control expressed in DuPont's 1938 policy statement, things would almost surely have been worse.

––––––

With peace and departure from Hanford, the DuPont corporate environmental staff returned to civilian work with far greater technical capacity and self-confidence than it had possessed before the war.[11] Its attention again turned to oversight of the company's operating divisions. The goal of creating a direct quantitative measure of environmental performance had proved impractical and was no longer pursued. In its place, there began in 1946 an annual survey of air and water pollution control at all of the company's plants, based on reports from the manufacturing divisions. A committee to coordinate the divisions' environmental activities was also established.

Initially, it seems, progress in pollution control at DuPont's plants resumed its gradual prewar pace. But the Donora calamity brought a new urgency. In presenting the January 1949 pollution survey to top management, the Engineering Department made the connection explicit. Pollution control, it noted, was getting more attention from regulators and legislators on local, state, and federal levels. The reason was public demand for stronger laws and better enforcement in the wake of the deaths at Donora and the weeks of national media coverage that followed. "The progress which industry as a whole has made," the Engineering Department observed, "has been generally unsatisfactory to those interested in accomplishing a clean-up."

The results of the company-wide survey were summarized in a chart showing the emissions from all of the firm's United States factories and the degree of government control over each. All but 14 of the 76 plants were located in "areas where governmental control agencies are active." Nearly half—34 of the 76—had been investigated by an enforcement agency.

The cost of the pollution abatement equipment already installed in DuPont's plants totaled $14,808,000. But this fell far short of what was needed—the Engineering Department's rough estimate of the expenditure needed to "comply with the intent of the laws on stream and atmospheric pollution" was an additional $20,500,000. The reason that achievement fell so far short was stated with clarity:

> ...there has been some reluctance on the part of the industrial departments to request appropriations to cover the installation of waste treating equipment, or recovery equipment not fully justified by savings, at existing plants.

To deal with this problem, the Engineering Department proposed a change in the 1938 trade waste policy. A more detailed and explicit statement was drafted for the Executive Committee, with two operative paragraphs. The first was a policy statement: at any company operation where pollution of air or water violated government regulations or otherwise created a nuisance, the problem would be promptly studied and the company would take what steps were necessary to abate the nuisance. The second had a more practical import. Investments in pollution control would be approved by the company's central management "on the basis of necessity," meaning that division managers would not have to justify them by showing a positive financial return. The staff recommendations took the company further than DuPont's Executive Committee was willing to go. The proposal was taken up at the March 23, 1949, meeting and was rejected. The 1938 policy statement was left in place, with the company president instructed to give a verbal reminder of its importance to divisional general managers.[12]

Under DuPont's decentralized structure, this was a crucial choice. The managers of each division were judged on the financial performance of their business; they had a strong incentive to forgo environmental projects that appeared only on the cost side of the ledger. No record of the Executive Committee's discussion of this matter has come down to us, so the rationale for the decision must be inferred. No doubt, cost was a concern; $20 million was nearly a quarter of the company's annual operating earnings, and executives surely suspected that the ultimate costs of an open-ended environmental commitment would exceed initial estimates.

But the main motivation may have been managerial rather than financial: an unwillingness to undermine the decentralized system of decision making that was essential to the efficient operation of a large and diverse company. At new plants, DuPont was willing to spend money on environmental control. Design and construction of new factories was the province of DuPont's central Engineering Department, the organization that housed the environmental staff. Proposals to invest in existing plants emerged from the manufacturing divisions. The Engineering Department's environmental staff, which could point to numerous divisions which would not undertake needed expenditures, lacked authority to make the line organizations put environmental protection ahead of short-term profitability. And top management was not willing to disturb the balance of power within the firm.[13]

Still, the Executive Committee's reaffirmation of the previous policy statement was not a decision to do nothing. Pollution remained on the front burner for DuPont management, with action continuing in several areas. The corporate environmental staff pursued the collection of information from the operating divisions, seeking a more detailed understanding of the company's emissions and the costs of cleaning them up. Under the combined pressure of regulatory oversight and internal suasion, the divisions moved to abate some of the worst problems. And in the design of new plants, DuPont sought to achieve a new and higher standard of emissions control.

———

In a company as large and diverse as DuPont, information was the key to central management's ability to exert any degree of control over far-flung operations.[14] During the first half of 1950, the manufacturing divisions submitted more detailed environmental reports that, for the first time, identified needed expenditures on an individual-plant basis. A summary report on air pollution was prepared by Carl Gosline of the Engineering Division on May 16; there was undoubtedly a companion report on water. The Executive Committee asked for further refinements, and on August 9 the Engineering Department submitted a three-page analysis signed by Chief Engineer G. M. Read, with four pages of accompanying tables. The company's total investment in pollution control, after items that really had other rationales were subtracted from the previous year's estimate, now came to $12,600,000.

The third and final page of the analysis submitted by the Engineering Department on August 9 concluded with an echo of the previous year's report:

It is evident that the Industrial Departments are making good progress in solving their pollution problems. It has been noted, however, that in spite of increasing pressure from the public and the regulatory agencies, the Departments are reluctant to request appropriations on the basis of necessity.

Such frankness was not to everyone's liking. The surviving copy of page 3 bears a penciled deletion mark. A revised page is substituted into the memorandum, with the sentence about reluctance to request appropriations omitted. The altered page is signed on Read's behalf by his deputy, M. F. Wood, with the handwritten August 15 date of the Executive Committee meeting.[15]

———

While policy was debated in the executive suites, the divisions began to deal with their problems. One obvious issue, after Donora, was the aging zinc plants DuPont had acquired in 1928 when it purchased Grasselli Chemical Company. What DuPont had sought in Grasselli, which remained an operating division within the larger company, was its strong position in the manufacturing of inorganic chemicals, including sulfuric acid and pigments. Zinc was part of the package; it was used to make lithopone, a paint ingredient that was one of the new division's main products.[16]

The zinc-making operations that U.S. Steel conducted at a single plant in Donora were divided between two locations by Grasselli. Ore was roasted, yielding sulfuric acid as a by-product, at New Castle, Pennsylvania, north of Pittsburgh. The roasted ore was shipped to a smelting plant at Meadowbrook, located upstream from Donora on a West Virginia tributary of the Monongahela. The Meadowbrook zinc plant was more modern than Donora's—vertical retorts, then a new technology, had been installed soon after DuPont's acquisition of Grasselli—but it stood out among the company's facilities in its need for environmental investment. The 1950 survey showed that $325,000 in air pollution controls were required at a fully depreciated plant whose original cost was only five times that amount.[17] And zinc was one of DuPont's least profitable businesses. Lithopone's share of the paint market had slipped during the 1930s, and zinc was a commodity product whose price was too low, except when wartime inflated demand, to earn the high return on investment that DuPont demanded from its managers.[18]

A policy decision to sell the zinc business was made in 1943 by DuPont's Executive Committee. But implementation lagged; it is not clear whether the Meadowbrook plant was kept on due to lack of buyer interest or the Grasselli Department dragged its feet on disposing of an operation that continued to yield some profits. In any event, Donora and the prospect of an expensive pollution control retrofit seem to have motivated the division to get serious about leaving the zinc business. On June 15, one month after completion of the air pollution survey, an agreement was reached to sell the zinc business to Matthiessen and Hegeler, a long-established but relatively small company that specialized in zinc. The sale price for Meadowbrook was $1.1 million. Final approval of the zinc disposition was considered by DuPont's Executive Committee at the same August 15 meeting that received the company-wide environmental survey. The relationship of the two matters was recognized, if somewhat obliquely, in a memorandum summarizing the transaction for top management. "It is believed," Grasselli's general manager observed,

> that if anything the Matthiessen and Hegeler Company will operate the New Castle and Meadowbrook plants with somewhat higher prof-its than du Pont because of probable lower overhead and because their operating standards are probably not as high as du Pont's.[19]

Another clear problem was the Electrochemical Department's facility in El Monte, an industrial suburb of Los Angeles. According to the 1949 survey, this plant was the only DuPont operation whose liquid effluent was in violation of regulations because of its toxicity. The toxic effluent was not discharged to a stream, but soaked directly into sandy ground. It is easy to see how this operation, conducted on the bank of the Rio Hondo two miles upstream from the sewer discharge that four years earlier had been the cause of the Montebello Incident, must have attracted regulatory attention. By the time of the next year's report, El Monte had disappeared from the roster of production facilities.[20]

Where outside pressure was lacking, however, the divisions often resisted expenditures that the corporate environmental staff thought necessary. A memorandum prepared before the August 15, 1950, Executive Committee meeting gave a half dozen examples from several divisions. The Rayon Department declined to install a $600,000 settling pond to remove floating solids from one plant's discharge. The Chambers Works, unwilling to pay for neutralization tanks, instead dumped lime directly from hopper-bottomed railcars into ditches. Yet another example involved the discharge of copperas and waste acid:

> Admittedly, in this instance the indicated investment was sufficiently high so that there was real justification in seeking more economical methods of waste treatment. At Baltimore a study is being financed by the Pigments Department to determine the effect of waste on the water of Baltimore Harbor. This, in part, is an attempt to develop

factual information and is in part a delaying action to avoid requesting the appropriation of money which would be required to neutralize acid and dispose of sludge and copperas by barging to the ocean.[21]

———

When DuPont built new plants, the Engineering Department exerted a direct control that it lacked when seeking retrofits to existing operations. The department's stated goal was to make DuPont's facilities models of careful attention to waste disposal. Setting out in 1949 to design a new plant in Victoria, Texas, to make adiponitrile, an ingredient in the recipe for nylon, it faced a more difficult task than usual. The Guadalupe River had a low flow rate, limiting its capacity to dilute wastes dumped into the stream. The company's response to this challenge involved both a thorough exploitation of known technologies and an innovation: the deep-well injection of hazardous chemical wastes.

DuPont began by commissioning a comprehensive inventory of the animals living in the river—a pioneering example of what is now called a biota survey. Two complete baseline surveys were carried out by the Philadelphia Academy of Natural Sciences, a year apart, and that work was repeated 18 months after the plant opened. At the mouth of the river, where shellfish were an important resource, the inventory was taken every six months.

The plant's main problem was getting rid of its liquid wastes. The most troublesome of these were the effluents from the production processes, which fell into three categories: organic liquids, dilute chemical solutions in water, and concentrated chemical brines. The organic liquids, along with combustible waste gases from the production process, were burned in a high-temperature incinerator with a 175-foot stack.

Dilute solutions were spread out in a lagoon to evaporate in the hot, dry Texas weather. The obvious danger was that these wastes would soak into the soil and contaminate groundwater. This was thought to be unlikely due to the low soil permeability at Victoria, but confirmation was desired. Test holes were drilled, and a layer of clay was found underneath the lagoon site. In addition, several monitoring wells were installed around the lagoon and the groundwater was tested periodically. Still, DuPont seems to have considered lagoon evaporation to be a somewhat dubious technique—the terminology of lagoon evaporation was often used by others as a euphemism for disposal into the ground. Two years after the Victoria plant opened, alternative disposal methods were under consideration.

The most difficult challenge was posed by the concentrated chemical wastes. These briny waters, containing 18% salt and more than 1% of organic chemicals, flowed from the plant at the rate of 80,000 pounds per hour. To dispose of them, DuPont adapted a technique that the Texas oil industry had long used to get rid of salty brines that come up with crude oil from their wells. Oilfield brines are commonly injected back into the geological formation they came from. Chemical companies had begun to experiment with putting concentrated wastes into rock layers filled with saline waters like the oilfield reservoirs that receive reinjected brines. DuPont decided to apply this technique in Victoria, where such a reservoir was available, located nearly a mile underground.[22] The deep-injection technique ran into delays that took several years to work out—the wastes, if not first treated with chemicals, reacted with underground waters and clogged the injection wells—but soon deep-well injection became a mainstay of DuPont's waste disposal practice. The technique was quickly adopted by other chemical companies. Dow, which had similar briny waste streams, made a decision for deep-well injection in 1951, and by 1953 it was fighting the same plugging problems as DuPont.[23]

Without a doubt, Lammot du Pont and other top executives wanted to do a better job of protecting the environment. Their company's high profit margins—its strategy was to seek out markets in which it could create a dominant position through some combination of patent protection, superior production technology, control of raw materials, or sheer market power—gave them room to do so. But DuPont's managers were trapped between the ideological imperatives of their political beliefs and the practical difficulties of running a large organization so as to maximize profits. The company's structure placed most decision-making authority in the hands of divisional executives. These men were judged by their ability to make profits and develop new markets. Inherent in this structure was an incentive to save money by skimping on pollution controls, an incentive that could only be overcome by direct supervision. Central management feared that the wildly successful decentralized structure could be undermined and was unwilling, except in rare circumstances, to override the divisions' authority.

Had the federal government imposed uniform rules on the entire industry, decentralization and economic competitiveness could have coexisted. This was an option that DuPont refused to entertain. The company was philosophically opposed to all forms of government regulation, and it was unwilling to make the same exception for the environment that it made, with profits directly at stake, for tariffs. DuPont insisted that industry must solve its own problems in its own way.

These two choices put a ceiling on what would be accomplished. DuPont's leaders would not cede to the government the authority to determine what their plants would emit, and they would exercise that authority only sporadically over their subordinates. Notwithstanding the continued nudging of the Engineering Department staff, and occasional showcase projects like Victoria, pollution control at DuPont continued to fall short of the goal the company had set for itself. In 1969, Samuel Lenher, who for the previous 14 years had overseen the environmental staff as vice president for engineering, took a backward look. He told a meeting of divisional environmental managers that

> In 1938, our company adopted a policy that pollution must be given equal attention with safety and fire protection. I must confess that I do not believe that we have met the challenge.[24]

CHAPTER 13 | The Industry Responds

We have to tell management that unless they do something, stringent laws will be passed which will require them to do something.

—Thomas Powers, Dow Chemical, 1954[1]

WHILE DUPONT AND OTHER individual chemical companies reacted to their own pollution problems, their industry also responded collectively. The industry's overall strategy, deeply embedded in the political worldview of its leaders, was clear: it would solve its own problems. Government regulation must sometimes be accepted, but with as little extent and effectiveness as feasible, and with local or state oversight preferred to national controls. To the greatest degree possible, what would be done would be determined by the industry on its own.

The industry's chosen means of action was its trade association, the Manufacturing Chemists' Association. The MCA operated largely through committees, each composed of a dozen or so representatives from member firms. The committee members had other work within their companies, and what was needed by one's own employer naturally took priority over a cooperative effort for the industry's common benefit, so in practice the work of these committees had an element of volunteer work. The committees were most effective when the enthusiasm of their individual members coincided with the economic and political needs of the corporations that employed them.

Distracted in its early years by the focus on tariffs, and hobbled by the limitations of its committee structure, the association at first did little about pollution. It intervened briefly in the oil pollution debate of 1924, but concerted action came only when the minds of top executives were concentrated by an imminent threat of government regulation. A water pollution committee was established in the spring of 1936; its immediate purpose was to formulate an industry response to Senator Lonergan's clean water bill. The air pollution committee followed only in 1949, a consequence of the alarm created by the Donora disaster.

———

The MCA's Stream Pollution Abatement Committee held its first meeting in May 1936, four months after the introduction of the Lonergan bill and a few days before congressional hearings were scheduled to begin. It was chaired by Walter Landis, a vice president of American Cyanamid. The four other members included representatives of Dow and DuPont. In its program as in its structure, the new committee followed the approach developed by the American Petroleum Institute after the passage of the Oil Pollution Control Act of 1924. Self-regulation was the cure for a problem whose reality could not be denied. The committee acted on two fronts; it undertook to share the companies' technical knowledge and set voluntary standards of environmental performance as it fought to stave off outside regulation.

Starting work in the shadow of the Lonergan bill, the chemical industry committee was at first concerned much more with political than technical matters. This focus is illustrated by the selection of Frederick "Zip" Zeisberg as DuPont's representative. Zeisberg had to his credit significant innovations in DuPont's production technology, but he was picked over another engineer whose experience was in the pollution field; his selection seems to have had less to do with his engineering competence than with his ardent belief in states' rights. After hearing the Lonergan bill discussed at a meeting of the Delaware Fish & Game Protective Association, Zeisberg wrote to a colleague that

> This bill was drafted by a New Dealer and under the guise of accomplishing a cleaning up of the streams of this country, which every right thinking man agrees should be done within reason, would take away all states' rights in this matter and center the control of virtually every stream and watercourse in this country in Washington.... Each state must make up its own mind just how far it wants to go in incurring the expense of cleaning up its streams. In doing this, it must consider the cost of cleaning up and balance this against the advantages obtained. There are unquestionably some streams in the country which might as well be recognized as sewers and used for that purpose only.

Landis wrote to Lammot du Pont that "it will be a great pleasure to have [Zeisberg] assist in the development of this project along sane and sensible lines."[2] What the industry thought sane and sensible, of course, was to be left alone by government as much as possible.

The industry continued after the war to follow water pollution issues closely at both federal and state levels. The MCA had a nationwide system for tracking proposed state water pollution laws, and it drafted a model state statute. The main focus of the stream pollution committee shifted, however, from legislation to technology, and in 1949 its name was changed to Water Pollution Abatement Committee "in view of the inclusion of ground waters in Committee activities."

The committee sponsored a pair of two-day technical conferences in 1947 and 1948. These well-attended meetings, like the American Chemical Society's 1946 symposium, focused mostly on the nuts and bolts of water treatment. The committee also tried to imitate the API by preparing manuals of good practice, but this task was hindered by the great diversity of manufacturing processes in the chemical industry. Only a single manual, giving general principles for the investigation of chemical plant wastes, had been issued by 1948.[3]

––––––––

For air pollution as for water pollution, it took the threat of outside interference to get the Manufacturing Chemists' Association moving. The association voted in 1947 to establish an Air Pollution Abatement Committee, but little was done in the ensuing year to put that decision into action. Only when the threat of government regulation loomed suddenly larger after the Donora smog disaster did the trade association spring into action. The medical director of Alcoa, one of the MCA member companies whose business straddled the boundary between nonferrous metals and chemicals, would make the connection clear at the new committee's second meeting:

> Present evaluation of the public health aspects of air pollution should give consideration to a change in the attitude of the public in this regard. The repercussions of the Gauley Tunnel episode on silicosis probably will be dwarfed by the effects of Donora on air pollution.

In January 1949, two months after Donora, the MCA issued a renewed call to member companies to nominate representatives to the air committee, and now interest was keen. An organizational meeting in March was quickly followed up with a technical conference on April 19 and 20 attended by 112 representatives from 51 companies.

The new Air Pollution Abatement Committee was constituted by the MCA's Board of Directors on September 13 and met on November 9.

Modeling itself on the water pollution committee, it took on a broad agenda ranging from the exchange of technical information to drafting industry positions on federal and state legislation. The members were well aware of the shortcomings of the chemical industry's efforts to control its air emissions, and they turned different faces to the outside and inside of the industry. Looking inward, they sought improved practices, realizing that this required not merely research and technological progress, but also what was delicately referred to as the "education of management." In dealing with the outside world, however, the committee pursued management's agenda of combating regulatory interference. Public statements minimized the dangers of air pollution to health. Regulation of air emissions by state and federal governments was opposed; in its place the committee promoted control by local governments, along lines laid out in what it promoted as "a rational approach to air pollution regulation."[4]

The contradictions inherent in these dueling objectives did not dissuade the committee from a vigorous program of activity. The two moving spirits were George Best of Mutual Chemical and the DuPont air pollution specialist Carl Gosline.

Carl Gosline, DuPont meteorologist who edited the air pollution manual of the Manufacturing Chemists' Association, photographed in the 1940s. (Courtesy of Anne Barnes.)

George Best, leading figure in pollution control activities of the Manufacturing Chemists' Association from 1949 to 1978. Shown in retirement in 1990, bringing flowers to shut-ins on Easter. (Photo by J. David Zimmerman.)

Carl Gosline (1920–97) joined the corporate environmental staff in DuPont's Engineering Department after working on the Manhattan Project at Hanford. Like Zip Zeisberg, DuPont's initial representative on the MCA water pollution committee, Gosline seems to have been chosen for his commitment to DuPont's political philosophy as well as for his technical skills. Relatives remember him as someone who "made Rush Limbaugh look like a liberal." Yet as a conscientious engineer—and a man who, although he never resided in one place long enough to put down deep roots, sought out involvement in each community where he lived—his attitude toward pollution control was anything but the denial of a Rush Limbaugh. At a 1952 conference he warned the public against the threat posed by wastes with unknown characteristics produced by new chemical processes.[5]

George Best (c. 1913–90), also among the initial members of the committee, represented a much smaller company than Gosline, yet he came to play an even greater role in the industry's work. Best received a chemistry degree from M.I.T. in 1934 and then spent nine years at New Jersey Zinc, a company which had been forced to install air pollution controls after litigation in the early 1920s. He was hired by Omar Tarr in 1948 to design the air pollution controls for Mutual's new Baltimore plant.

Although Best was Tarr's junior by two decades, the two men had remarkably similar backgrounds that must have shaped their outlook on the responsibilities of engineers. Both were natives of Maine who married wives from northeastern Pennsylvania. Both were elders of Reformed churches; as a student, Best had been president of M.I.T.'s Christian Association. They were alike as well in their recognition of their industry's environmental failings and in their commitment to correcting them from the inside. Where they differed was in personality. Tarr carried the engineer's emphasis on hard data and insistence on scientific precision into his relations with others. Best too was very much an engineer, but he was a modest man with a talent for consensus building.[6]

Information exchange was the first activity of the new committee to get off the ground. Technical information was the main concern, but legislation and public relations were not ignored. Gosline was appointed to oversee the production of a pollution abatement manual, and a second industry conference was scheduled for the following April.[7]

In resolving to write a manual, the committee followed the lead of the water pollution committee, but it went far beyond the fragmentary efforts of its sister body and produced a systematic survey of the field. Carl Gosline oversaw the entire effort, recruiting authors, prodding reviewers, personally editing the text, and seeing the project through to a conclusion even after he left the committee. The fruit of his labors was a 400-page document that covered a complete range of topics from public relations and legislation to monitoring and control technologies. Each of 12 chapters was drafted by a different specialist, most of them employees of member companies. The drafts were circulated to the entire committee for comments and revised before receiving final approval. Each chapter was printed as a separate pamphlet when ready, starting in 1951. Gosline continued his editorial work even after resigning from the committee in late 1952 upon his reassignment to the new plutonium production plant in South Carolina. A replacement editor was finally named after the last chapter appeared in 1954, but with Gosline's departure the impetus was lost; a plan to issue regular revisions was never carried out.[8]

The task initially apportioned to George Best was the exchange of technical information, but with others taking charge of the manual and the annual conferences, this ill-defined assignment gave him little to do. At a September 1950 meeting, Best jumped into a discussion of legislative proposals. Five principles had been drafted by Charles Caspari of Monsanto, the chief of which were that "The atmosphere should be regarded as a useful natural resource...utilizable for dispersion of wastes within its capacity to do so without harm to the surroundings" and that "Air pollution is a local problem." Best suggested that it would also be useful to have state agencies that would provide technical assistance to local enforcement personnel. He followed this idea up a month later in a written memorandum that the committee received with enthusiasm. In March, a subcommittee consisting of Best, Gosline (who was representing the MCA in the drafting of state legislation in New Jersey), Dudley Irwin and Frank Seamans of Alcoa, and C.H. Bunn of Standard Oil was appointed to draft an industry position paper. With Caspari having absented himself since the previous September's meeting, Best was chosen to chair the group.

A draft of what was called "A Rational Approach to Air Pollution Legislation" emerged in May; additional review and revision followed at several meetings during the remainder of the year. Best presented the committee's product as a paper to the Industrial Hygiene Foundation in November and to the MCA's air pollution conference the following February. The underlying concept of treating the atmosphere as a sink for wastes was unchanged from Caspari's original principles, but Best added an organizational structure resembling the water boards set up two years earlier under California's Dickey Act. Standard setting and enforcement would be in the hands of local commissions, on which representatives of industry would have a majority. A state agency would provide technical support but have no direct authority to regulate.[9]

The MCA committee did not confine itself to stating principles of regulation; it concerned itself with the implementation as well. It formulated the industry's reaction to model laws written by technical societies and discussed both substance and strategy with regard to bills under consideration in individual states and localities.

Notwithstanding the rhetoric about a "rational approach,"—to which Best, at least, was sincerely committed—the MCA in practice sacrificed consistency to the goal of avoiding government control. It sometimes opposed standards for being insufficiently specific; on other occasions it criticized proposed rules for singling out specific chemicals. When a committee of the Air Pollution and Smoke Prevention Association of America proposed national standards for air pollution, industry representatives objected that

> The standard for emissions in the majority report makes no allowance for the tremendously varied kind and varied toxicity of dusts and fumes emitted by industry. A plant crushing rock for road materials and emitting relatively non-toxic dusts is required to provide the same degree of controls as a plant smelting lead battery plates and emitting lead dust and fume.[10]

On the other hand, a Utah bill was "extremely objectionable" when it was so specific as to declare concentrations above 1 ppm of SO_2 or arsenic in the air above property of others a public nuisance subject to abatement.[11]

The effort to avoid regulation involved public relations as well as lobbying. At the second meeting of the air pollution committee, the objectives for the association's public relations effort were set out in a straightforward way:

(1) to offset the adverse effects caused by the activities of irresponsible headline hunters and troublemakers, (2) to prevent the development of public demand for drastic and impractical air pollution and smoke control legislation, and (3) to educate the public as to the difficulty of eliminating and controlling air pollution and what the chemical industry is doing about it in order to gain member companies the time necessary to solve their problems in the most practical manner.

The minutes then report that

It was agreed that the chemical industry should not be over-zealous in this regard for fear that it might be singled out as the principle [sic] contributor.[12]

The air pollution committee recommended that MCA create a public relations committee. Such a body had actually been established on a modest basis several years earlier, and the two committees were soon working together. At the instance of the public relations specialists, the association's annual pollution conferences were opened to the press. At these public events, as elsewhere, the industry took the stance that air pollution was "not a serious or critical menace to public health." This was in distinct contrast to the advice it had received privately. At the very meeting where the public relations campaign was first outlined, Alcoa's medical director, Dudley Irwin, had reported: "As so little factual information exists with regard to the acute or chronic affects of general air pollution, a considerable diversity of honest opinion will probably persist for some years to come."[13]

Having completed the development of an industry position on air pollution regulation and with means of promoting that position in place, the air pollution committee sought new frontiers. A Project and Program Subcommittee was established to chart its course, with George Best, his work on the "rational approach" complete, as chair. This group soon returned to the early concern, not as yet followed up, about the education of the industry's own management. Ed Adams of Dow wrote to Best, suggesting that the committee concern itself with the "attitude of industrial management." As he explained,

I am apprehensive that too many persons of this group are not sufficiently concerned about pollution and are not adequately informed of the nature of the problem or of technical matters. Perhaps there isn't much to concern us here; on the other hand I have encountered rather frequently in the past, reports which cause me to wonder.

Any attack falls into the broad category of propaganda. I would propose a series of "case histories," accounts of situations and recent developments, of incidents, and corrective measures, to be mailed to individuals and to be published in the trade news magazines.

When the subcommittee presented its report a year later, education of industrial management was the first and most discussed point. Best "felt that management needed to develop a conscious acceptance of a responsible attitude toward air pollution control."[14] Thomas Powers, a long-time Dow water pollution specialist who was filling in for Adams, expressed an even stronger view. He argued that management was fully aware of the problem and needed to be told "that unless they do something, stringent laws will be passed which will require them to do something." Powers, who had been a state regulator before he joined Dow, was not averse to helping that process along. Two years earlier, representing industry on the American Water Works Association's committee on groundwater pollution, he signed on to a report that described that problem as having "rather wide distribution" and called for stronger regulation.[15]

George Best's work on the air pollution committee soon came to an end. After Mutual Chemical was sold to the larger Allied Chemical in 1954, he was not renominated for membership. In 1957, the Mutual Chemical technical staff was moved to Syracuse. Two years thereafter, Best left Allied and returned to the committee work that had so clearly engaged him, becoming a full-time MCA employee in charge of the association's work on air and water pollution. He would remain at MCA, rising to the position of secretary-treasurer, until his retirement in 1978.[16]

————

The program of environmental self-regulation that the chemical industry adopted in the 1930s has an inherent weakness. Price competition always favors the producers with the lowest costs. Unless outside compulsion forces all to abide by the same standards, responsible environmental behavior will at best reduce profits, and at worst can drive a company out of the market.

In the 1920s, when the oil industry developed the concept of self-regulation, and in the mid-1930s, when the chemical industry adopted it, the dangers of price competition were largely theoretical. There was little of it in either the oil or the chemical business. In domestic chemical markets, the lax antitrust enforcement of the Republican administrations of the 1920s tolerated informal understandings by which manufacturers agreed not to poach on each others' customers, while world markets were controlled by formal cartels organized by the dominant German producers. These arrangements were not threatened by the early New Deal, which viewed price-cutting as a destructive force and established the National Recovery Administration to control it.

But antitrust enforcement resumed with a vengeance when the New Deal turned left during Roosevelt's second term, and a series of lawsuits were brought against major chemical manufacturers. In May 1941, eight companies, including DuPont, Allied Chemical, and American Cyanamid, were indicted for conspiring to monopolize the dye industry. Antitrust enforcement continued to target the chemical industry after the Second World War, culminating in the Justice Department's successful effort to separate General Motors from DuPont. Meanwhile, the growing integration of the petroleum and chemical industries and the sale of government-built plants after the war pushed new competitors into markets. No longer could a small club of producers avoid price-cutting with a wink and a nod.[17]

Competition might have come to the markets where chemicals were sold, but monopoly power still held sway in the marketplace of environmental ideas. The ideological predispositions of industry leaders were reinforced by the difficulty of reversing course in a trade association that operates by consensus, and the industry maintained its unyielding and highly effective opposition to outside control. With new legal and logistical obstacles making it hard for companies to collude in setting prices, increasingly severe price competition inevitably undermined self-regulation.

A DuPont, whose technological prowess and market power allowed it to focus on high-margin businesses, still had room to meet the standards that industry leaders articulated. Even so, it often fell short. The tools of persuasion, information, and statements of policy that central management wielded were, as we have seen, insufficient to induce its divisions to attain the company's stated goals. Many other manufacturers were in weaker positions. They made commodity products that became subject to fierce price competition. With market forces left unchecked as a result of the industry's still effective opposition to government regulation, makers of chemicals like zinc and DDT faced a hard choice. They could either operate with a higher cost structure than their competitors or allow themselves to be driven down toward the level of the dirtiest and least scrupulous.

In the absence of outside compulsion, spending on environmental cleanup remained modest. The MCA reported as late as 1954 that the industry's annual expenditures on pollution control were $40 million, a small sum compared to its $20 billion in sales and $3 billion in pretax profits.[18]

CHAPTER 14 | From Donora to Love Canal

When President Nixon and his staff walked into the White House on January 20,
1969, we were totally unprepared for the tidal wave of public opinion in favor of
cleaning up the environment that was about to engulf us.

—John Whitaker, 1988[1]

POLLUTION RARELY CONFERS FAME, and when it does it exacts a
high price. Two communities that suffered the misfortune of making
environmental history are Donora, Pennsylvania, and Love Canal in Niagara
Falls, New York. The two towns are markers in space, delineating the heart-
land of America's industrial revolution, as the catastrophes they suffered are
bookends in time. The crow's flight between them of barely two hundred
miles traverses the cradles of two industries, the steel center of Pittsburgh
and the oil fields of Titusville. The two disasters serve as markers in history,
bracketing the journey from the apogee of industry's fight against regulation
to the era of government control.

———

The Donora catastrophe of 1948 made it clear that factory fumes were more
than just another species of smoke. The Manufacturing Chemists' Association was not alone in bringing new vigor to pollution control. Regulatory
initiatives were launched in New Jersey and other industrial states; in a sign
of the times, the Smoke Prevention Association, composed mostly of local

regulators, added Air Pollution to its name in 1950. Industrial air pollution now had a place on the national agenda, from which it would never again be dislodged.[2]

What was to be done about the newly urgent hazard? The Industrial Hygiene Division promoted its usual remedy: more research. But the experts who had been squeezed out of the Donora study—the Bureau of Mines, the Steelworkers union, and independent scientists—were not silenced. The Bureau of Mines in particular grew more vocal about the need to clean up the nation's air. Louis McCabe, now heading its environmental programs, urged a federal air pollution program that went beyond research to control.

Industry and its allies still sought to limit the federal role. Within a year, Robert Kehoe was lobbying to have the Bureau of Mines' air pollution program ended and the field made the exclusive preserve of the Public Health Service. The two agencies competed quietly for control of air pollution programs for another five years.[3]

Public worry about the air was propelled by new pollution episodes. A week of killer smog hit London in December 1952, more than doubling the death rate until it cleared.[4] In 1954, Republican senators from the smog-afflicted states of California and Indiana submitted legislation to establish a federal program of air pollution research; the Eisenhower administration responded by undertaking to draft a bill of its own.

The chemical industry would have preferred no federal action at all, but it could not oppose a program limited to study. It endorsed the concept of a federal research program and used the legislation as a vehicle to get the Bureau of Mines out of the air pollution business. On April 20, 1955, the president of the Manufacturing Chemists' Association complained in a well-publicized speech about "duplication of effort and friction among various agencies and organizations in research and preventive measures." Ten days later, Louis McCabe transferred into the Public Health Service. In June, the administration submitted its bill, which passed quickly and without open controversy. The Air Pollution Control Act of 1955 strictly limited the federal role in keeping the atmosphere clean. The national government's only task was research, to be carried out by a new arm of the Public Health Service. The Bureau of Mines was out of the picture, relegated to a seat on an advisory committee.[5]

The promise made on Louis McCabe's transfer, that he would oversee the government's air pollution program, must have been an empty one. The new federal air pollution research organization was still part of the Industrial Hygiene Division and retained its anti-regulatory bias. Although McCabe was far from retirement at the age of 51, he left government service after eight months, with no prospects beyond those of setting up his own consulting business. The remainder of his career was spent in the private sector.[6] His fight against air pollution had run up against the same obstacles that stymied control of pesticides and environmental cancer. Political influence had marginalized opinions that scientific argument could not refute.

––––––––

Pesticide safety was another environmental problem that remained in the public eye through the '50s. A special committee of the House of Representatives under Rep. James Delaney of New York, established in 1950, held 46 days of public hearings around the country. The Food and Drug Administration, still supporting the precautionary principle it had backed in 1930s debates over lead arsenate, wanted new pesticides to be allowed on the market only when proven safe. Extended negotiations over new legislation led to the Miller Amendment of 1954, which allowed the FDA to limit the uses of a pesticide on fresh produce but still denied it the power to keep a substance entirely off the market. In its clear rejection of the precautionary principle, the Miller Amendment is seen by historians as a victory for the pesticide manufacturers.[7]

By now, a close alliance had emerged between the Agriculture Department and the pesticide manufacturers. Ties were cemented with lobbying, political pressure, funding for research at university schools of agriculture, and hiring away government scientists to promote new products. The department no longer sought to strengthen the permissive rules established by the 1947 pesticide law and established its own spraying program. It consistently rejected claims that pesticides had injured humans or livestock.[8]

The Miller Amendment did not still the criticism of pesticides, and further controversy was triggered by aerial spraying campaigns. After renewed hearings in which the suppression of Wilhelm Hueper's research came to light, hard bargaining among the contending interests led in 1958 to agreement on a modest strengthening of pesticide controls. Delaney by this time was a senior member of the powerful Rules Committee, and he took advantage of the ability to block legislation that this position conferred. He inserted into the bill a clause whose broad import was poorly understood by his colleagues, forbidding any detectable residue of a carcinogen in processed foods. A year later, a national furor erupted when the FDA used this provision to condemn the entire cranberry crop. The action was announced a few weeks before Thanksgiving, the day on which 70% of the entire year's production was consumed.[9]

The next step in the pesticide debate—and a turning point in public attitudes toward chemical pollution—came in 1962 with the publication of

Rachel Carson, author of *Silent Spring* (1944 photo). (Courtesy U.S. Fish and Wildlife Service.)

Silent Spring. Rachel Carson, a gifted nature writer with an earlier best seller to her credit, gathered together the research of many scientists into a comprehensive critique of chemical pesticides. Echoing the entomologists who had warned two decades earlier of DDT's threat to the balance of nature, she portrayed an ecosystem under siege. Chemicals that killed insect pests also kill the predators that keep the pests naturally under control. As the pesticides move through the natural food chain, they are concentrated in the tissues of predators. Ubiquitous in the environment, they inevitably reach human beings, inflicting a toll of poisoning, genetic damage, and—here Hueper was quoted at length—cancer.

The June appearance of *Silent Spring* as three long articles in the *New Yorker* magazine struck a chord in public opinion. The articles were quickly expanded into a book, an immediate best seller that sold a half million copies within six months and remains in print today. Favorable reviews and sympathetic editorials flooded in from independent scientists and the nonscientific media. The *New York Times* editorial page suggested that Carson might turn out to be as deserving of a Nobel Prize as Paul Müller, the Swiss chemist who had discovered DDT's insecticidal powers. An hour-long television special was broadcast on CBS the next spring.

The chemical industry and its allies in government were quick to counterattack. Even before *Silent Spring* appeared as a book, the Manufacturing

Chemists' Association launched a campaign of mailings, press releases, and speeches. Individual companies and allied trade associations joined in, with Velsicol Chemical Corporation trying to squelch the book by threatening to sue the publisher. Industry-allied scientists at the Agriculture Department and elsewhere hit sharply at Carson as well. But the vehemence of industry's response, depicting Carson as not merely a critic of pesticides but an enemy of scientific progress itself, only magnified her influence.[10]

This massive explosion of public attention and scientific concern was not enough, in the short run, to overcome the entrenched political positions of the defenders of pesticides. In August 1962, President Kennedy asked the President's Science Advisory Committee to examine the issue. That body's report, issued the following May, came down on Carson's side. It called for a phaseout of DDT and a string of reforms in the 1948 pesticide act, including an end to "registration under protest" of compounds rejected by the regulators. A flurry of bills were submitted in Congress, but chemical industry lobbying convinced the House to water down a modest reform bill passed by the Senate, and conservative legislators succeeded in staving off change for the remainder of the decade.[11]

The edifice of scientific orthodoxy erected by Royd Sayers, Anthony Lanza, Robert Kehoe, and their collaborators was under assault from more than one direction. On one side was the wave of popular and media attention that followed *Silent Spring*. From another came advances in environmental science, knocking out the underpinnings of the scientific defenses that polluters had labored diligently to build.

The safety of leaded gasoline was questioned again as automobile use skyrocketed. One of the pillars of Robert Kehoe's defense of lead, closely linked to his assertion that exposure at low levels is harmless, was his explanation of his finding that the metal was ubiquitous in human blood. Kehoe asserted that the levels found in the general population came from the natural environment rather than pollution. The geochemist Clair Patterson, who had made his reputation by using precise measurements of lead isotopes to make the first accurate determination of the age of the earth, refuted this belief. Careful testing of buried ice from Greenland showed that even in that remote land, lead levels in the atmosphere rose sharply after leaded gasoline was introduced. In a widely publicized 1964 paper, Patterson went beyond his purely scientific findings. Reviving the precautionary approach to lead exposure, he asserted that "the average resident of the United States is being subjected to severe chronic lead insult." The lead industry and its academic supporters counterattacked furiously. By 1966,

Kehoe and Patterson were offering clashing opinions to a congressional hearing.[12]

The dangers of asbestos were at last exposed as well. While earlier publications had described associations between asbestos and cancer, the connection became widely accepted after a 1964 study by Irving Selikoff, director of a pioneering occupational and environmental health clinic at Mount Sinai Hospital. Selikoff had taken on the subject after noticing high cancer rates among asbestos workers treated at his clinic in New Jersey, and he campaigned passionately for control of asbestos hazards.[13]

With the tide of scientific opinion shifting, Wilhelm Hueper saw his long struggle for recognition of occupational cancer vindicated as he approached retirement. The United Nations awarded him a medal for cancer prevention in the year of *Silent Spring*'s publication, and further honors were bestowed in the years that followed. The National Cancer Institute—which had bypassed Hueper when the Kennedy administration resumed environmental cancer research in 1961, brushing off an extraordinary complaint from Rep. John Dingell that such an action reflected "nothing but discredit upon the National Institutes of Health and upon the Surgeon General"—had its own perestroika in the mid-1970s. An extraordinarily self-critical history of the institute was commissioned for its fortieth anniversary in 1977, and Hueper received the institute's Director's Award a few months before his death the following year.[14]

In the aftermath of *Silent Spring*, a new political landscape emerged. Fish kills, beach closings, and stinking rivers were ever more evident, and in Rachel Carson's interpretive framework they were not mere nuisances but a threat to the planet. The tools in the hands of the federal government were not remotely equal to such an emergency. When the Water Pollution Control Act was amended in 1956 and 1961 to fund construction of sewage treatment plants, lobbying by chemical manufacturers and other industrial interests had ensured that no meaningful controls were placed on industrial pollution.[15] Not until 1963 were any regulatory powers added to the air pollution law that had been passed in 1955, and these were modeled on the enforcement provisions of the earlier water pollution laws, whose lack of real value had long since been made clear.[16]

Senator Edmund Muskie of Maine now stepped forward to seek legislative expression for what was already more than an issue but not quite yet a movement. The Public Works Committee in 1963 established a subcommittee under his chairmanship to deal with air and water pollution. His first initiative was a water pollution bill, enacted in weaker form two years later.

Federal enforcement powers were strengthened, but not enough to give them any real effect, and the federal water pollution bureaucracy was removed from the industry-friendly Public Health Service.[17]

Muskie turned next to the air. He faced a severe obstacle in the Public Health Service's Air Pollution Division, which had been fully separated from the Industrial Hygiene Division only in 1960 and remained a bastion of resistance to any interference with industry's right to pollute. By now Secretary of Health, Education, and Welfare John Gardner, who was the division's nominal overseer, favored national regulation of air emissions. As a former governor, Muskie was less inclined to centralization and preferred to keep standard setting in local hands. The Clean Air Act of 1967 hewed close to Muskie's views, aiming to force genuine cleanup while retaining the system of local primacy.[18]

As the 1960s drew to a close, a rapid succession of pollution crises drew wide publicity. In January 1969, an oil well blew out off the California coast near Santa Barbara, spilling three million gallons of oil and killing

The burning Cuyahoga River. This 1952 photograph won national attention after the Cuyahoga caught fire again in 1969, helping to trigger the environmental movement of the 1970s. (Courtesy of the Cleveland State University Library, Cleveland Press Collection.)

thousands of birds and sea animals. Cleveland's Cuyahoga River caught fire in June; similar conflagrations in earlier decades had gained little notice outside the city, but this fire became a national scandal. The following March, chemical plants were found to have spread mercury pollution through wide areas of the Great Lakes; fishing was strictly limited in the polluted waters.[19]

Earth Day on April 22, 1970, turned environmental protection into a massive grass-roots movement. Sen. Gaylord Nelson of Wisconsin proposed a national teach-in, modeled on one of the most successful tactics of Vietnam War opponents. This event mushroomed into what may have been the largest of all the protest actions of the '60s. New York City's Fifth Avenue was closed to automobiles so a crowd of 100,000 could assemble; organizers claimed that 20 million people participated around the country. The membership of grass-roots environmental organizations, new and old, skyrocketed, and pollution control finally had an effective lobby in Washington.[20]

View of the crowd at an Earth Day rally in Philadelphia, April 22, 1970. (Courtesy Temple University Libraries, Urban Archives, Philadelphia, Pennsylvania.)

A confluence of circumstances now led a conservative president to preside over a burst of environmental progress that far outshone the accomplishments of Theodore Roosevelt, Franklin Roosevelt, and Lyndon Johnson. The complex personality that was Richard Nixon had many facets; one of them was a conservative reformer who modeled himself on the nineteenth-century British statesman Benjamin Disraeli. "Tory men and Liberal policies," the president remarked in 1969, "are what have changed the world."[21] In a country already riven by the unpopular Vietnam War, moreover, Nixon was loath to set himself against public sentiment on another emotional issue. Finally, and more practically, Edmund Muskie was favored to win the Democratic presidential nomination in 1972. No incumbent wants to hand his likely opponent a campaign issue. Nixon, then, would take a stand for the environment; the chemical industry and other interests opposed to federal pollution controls were left to fight a rearguard action against overwhelming popular and political pressure.

Two months after Earth Day, the president issued an executive order establishing a new regulatory agency, the Environmental Protection Agency. The new EPA was soon armed with the powers that had been denied to federal regulators two decades earlier.

In December, the month the EPA went into operation, Muskie's Clean Air Amendments of 1970 were signed into law. Muskie's views had been radicalized by the ineffectiveness of the 1967 air act, and the aggressive staff director of his subcommittee, Leon Billings, was uncowed by impressively credentialed defenders of the status quo. Their revolutionary 1970 statute established a comprehensive federal system of regulation. Emissions standards would be the same wherever a plant was located, so that polluters could not play one locality against another to force everyone down to the lowest common denominator. Day-to-day enforcement would remain in the hands of the states, but the federal government could step in if local oversight lagged. Moreover, if both state and federal authorities failed to meet their responsibilities, individual citizens could go to court.[22]

In that same month, the president issued another executive order establishing a permitting system for industrial pollution discharged into streams. For this purpose, Nixon invoked the Corps of Engineers' authority under the Rivers and Harbors Act of 1899, revived a few years earlier after lying dormant since Herbert Hoover faced down the Corps in the oil pollution debates of the 1920s.[23]

Muskie and Billings, their rewriting of air pollution law accomplished, undertook a similar revolution in water pollution control. Their Clean Water Act, passed unanimously by the Senate and weakened only slightly in the

Senator Edmund Muskie in 1971 with Don Nicoll, his chief of staff, center, and Leon Billings, his chief environmental aide. (Courtesy of The Edmund S. Muskie Archives and Special Collections Library, Bates College.)

House, upended the entire framework of regulation. A national permit system covered sewage as well as industrial waste and closed loopholes created by the insufficiencies of the 1899 law. Instead of protecting the uses of streams—and allowing pollution when other uses were not impeded—waters were now to be cleaned up everywhere. Uniform national standards would require plants to use the same treatment technology wherever they were located, so that factories could not repulse cleanup requirements by threatening to move to another state. Citizen suit provisions mirrored the Clean Air Act, to ensure that the new rules would be enforced. The final bill passed both houses as Congress prepared to adjourn in 1972.[24]

The political situation was now drastically changed, with Nixon headed for a landslide election victory over George McGovern. The president issued a veto on October 17. The veto message pointed to the bill's generous funding for local sewage treatment plants; objections from chemical companies and other manufacturing interests may have been more influential behind the scenes. But Congress had stayed in session expressly to guard against this, and the veto was overridden by an overwhelming vote.[25]

The third post–Second World War environmental law, the pesticide act passed in 1947, had to be dealt with as well. A rewriting of this statute, the Federal Insecticide, Fungicide, and Rodenticide Act, was inevitable given the temper of the times; DDT was already being phased out nationally after the accumulating pressure of public complaint, scientific criticism, publicity, and lawsuits led by 1969 to a wave of local restrictions. Nixon's new EPA proposed a relatively strong pesticide bill in February 1971. The legislation was taken up first in the House of Representatives, where chemical and agribusiness interests dominated. The bill passed by the House in November was so severely watered down that one senator could describe it as even weaker than the existing law.

Events took a different course in the Senate. The bill was vetted by Muskie's subcommittee alongside the Agriculture Committee, which normally had jurisdiction. Environmentalists gained new backing from labor unions increasingly worried by occupational disease, and two Democratic-leaning farmers' organizations defected from the farm bloc. A relatively strong bill emerged from the Senate as adjournment approached on September 26. At this point, the Nixon administration reversed course as it had on the Clean Water Act and lobbied the conference committee for the weaker House bill. The compromise that emerged on October 5, one day after the passage of the Clean Water Act, was more acceptable to industry than the new air and water statutes. Nixon signed it into law on October 21.[26]

———

The EPA's powers over air, surface water, and pesticides did not exhaust its mandate. From the beginning, the new agency was charged with another problem—the dumping of hazardous wastes and the groundwater pollution that often resulted. The problem of hazardous wastes burst into the public arena in the summer of 1970, when the Army got rid of outdated poison gas in what it called Operation CHASE—an acronym for "cut holes and sink 'em." Some 12,000 warheads filled with nerve gas were shipped across several states and dumped into the Atlantic Ocean. Coming a few months after Earth Day, this story was a natural for television; it made the three networks' evening news programs 39 times in a single month. Congress responded immediately, and in legislation enacted that October the new agency was instructed to plan a national system of disposal sites for hazardous wastes.

The need was urgent; a safe outlet had to be found for the dangerous substances that were no longer to be spewed into the air and water. But regulators lacked the legal authority—and even more, they lacked the knowledge—that they needed before they could finish the job of controlling industrial effluents.[27]

The next few years saw a welter of conflicting legislative proposals for dealing with toxic wastes. Much of the debate focused on a family of toxic synthetic organic compounds, PCBs, that in the late 1960s had been found to contaminate ecosystems worldwide. Intense negotiations involving industry and environmental lobbyists led to passage of the Toxic Substances Control Act of 1976. The manufacture of new PCBs was forbidden, and EPA gained authority to control particularly risky chemicals at any point between the manufacture of a product and its disposal.[28]

The Toxic Substances Control Act, as things turned out, had little impact beyond the PCB ban. The powers it granted were fettered by so many restrictions that it was nearly impossible to exercise them. But another law, enacted quietly in the same year, established what has become the third pillar of EPA's regulatory authority, the control of hazardous wastes. This was the Resource Conservation and Recovery Act, or RCRA.

The advertised purpose of RCRA, as its name suggests, was to encourage recycling of waste materials. The 1976 legislation amended what had started out as the Solid Waste Act of 1965, a law similar in spirit to the 1948 water pollution law and the 1955 air pollution law that provided for research and technical assistance. What public debate it did engender focused on rules that would have required deposits on soft-drink bottles and cans. The far-reaching hazardous waste provisions were inconspicuously tucked away in a brief passage. The bill was hustled through Congress in the preadjournment rush and passed the Senate only on the last day of the session. As one environmental lobbyist, Blake Early of Environmental Action, later recounted, the quiet was a deliberate stratagem:

> We pursued an intentional "stealth" strategy of focusing public attention on the need to adopt the "bottle bill"...as a means of diverting industry attention away from the hazardous waste provisions. We were certainly cognizant of the fact that TSCA was also holding their attention as well.[29]

————

What would turn out to be among the most troublesome of all the hazardous pollutants were in 1970 still a small blip on the radar screen of government regulators and the public. From the time of Lyne and McLachlan's 1949 report of TCE in water, the chlorinated solvents were known as groundwater contaminants, listed in published surveys among many substances that merited concern but not thought worthy of special mention by government scientists.[30]

The presence of chlorinated solvents in drinking water came to new attention just a few months after EPA was formed. A Senate hearing in New Orleans in April 1971 was told of a long list of chemicals found in the city's odorous water supply, drawn from the Mississippi River downstream of a vast complex of petrochemical works. The substances detected included PCE and many other chlorinated compounds.[31]

Meanwhile, an Italian scientist had found cancers among rats exposed to vinyl chloride, a compound closely related to PCE and TCE. Vinyl chloride was a commercially important product to which the public was exposed in hair sprays and elsewhere. It is also the raw material for the widely used plastic PVC. European chemical manufacturers quickly commissioned a study by another Italian, Cesare Maltoni. By the fall of 1972, Maltoni had discovered that vinyl chloride was an exceptionally potent carcinogen. American chemical companies were quickly informed, but the result was not disclosed to the public.

The Manufacturing Chemists' Association decided to undertake its own study. George Best, by now in charge of all of the MCA's environmental and safety activities, still retained the belief in manufacturers' responsibilities that he had shared with Omar Tarr. He proposed involving the government in the research—which would necessitate the disclosure of Maltoni's data. The industry, he argued, would be better off getting out in front of bad news. He then commented that if the additional work confirmed the earlier findings there was a "moral obligation" to disclose the danger. The industry did not rush to take his advice, and the experiments became public only after employees of a Kentucky vinyl chloride plant were diagnosed the next year with the same rare type of cancer observed in Maltoni's rats.[32]

The vinyl chloride discoveries touched off an immediate uproar about environmental cancer. The contamination found in New Orleans drinking water took on a new dimension when an Environmental Defense Fund study found high rates of cancer in communities that draw their water from the Mississippi.[33] In the Safe Drinking Water Act passed in December 1974, Congress ordered a comprehensive study of carcinogens in drinking water, to be completed in six months. EPA responded to these developments by following up on the New Orleans study with a survey of organic chemicals in the water supply of 80 cities. For this purpose, EPA needed to improve its analytical techniques; laboratories of the day were often very adept at measuring specific chemicals that were identified in advance as targets, but methods that could scan for a wide spectrum of organic compounds were not well developed.[34]

The new lab techniques, which could detect chemicals that weren't being looked for, were quickly adopted by other public health agencies. Among these was the New York State Health Department. At just this time, that agency was being called on to investigate complaints about the taste and smell of water at Grumman Aircraft on Long Island, the same plant where in

1943 yellow drinking water had turned out to be contaminated with chromium. With the new technology, the chemicals in the groundwater were identified—they included vinyl chloride, TCE, and PCE. Two Grumman wells were ordered shut in August 1976.[35]

The discovery of a known carcinogen triggered a countywide survey of drinking water. In the small city of Glen Cove, widespread contamination by chlorinated solvents turned up in public supply wells. By 1978, nearly 400 wells in Glen Cove had been tested and TCE and PCE—by then also identified as carcinogens—and other solvents had been found at concentrations ranging from ten to the hundreds of parts per billion.[36] Similar solvent contamination was turning up elsewhere. After a May 1976 survey found cancer hot spots like Louisiana's, New Jersey began a statewide survey of drinking-water wells. In March 1978, a report was issued showing high levels of chlorinated solvents in many parts of the state.[37] A month later, the prominent groundwater scientist David Miller told a congressional hearing in Glen Cove that TCE and similar solvents "commonly migrate in ground water and are even more important than some of the pesticides."[38]

———

As government was at last bringing under control the pollution of air, water, and earth, there emerged yet another environmental problem, one that might dwarf others in its worldwide effects. This was the threat that the growth of industrial civilization would lead to global warming.

Global warming was not a new idea. The mechanism of the "greenhouse effect"—carbon dioxide in the atmosphere traps the sun's heat and helps keep the earth warm—had long been known. Some of the greatest nineteenth-century scientists discussed whether humans would warm the earth by burning coal. In 1908, a widely read book by the renowned Swedish chemist Svante Arrhenius predicted that the effect might show up within a few centuries.

But these early speculations met strong objections. Water vapor also traps heat, and it wasn't clear whether the warming effect of carbon dioxide would really add anything to what the moisture in the air was already doing. Would the carbon dioxide emitted by burning fuels collect in the atmosphere, or might it not quickly dissolve into the ocean or be taken up by plants? Would factory smoke block out heat and cause the earth to cool rather than warm? The workings of the atmosphere were imperfectly understood, and even when the basic principles were known, the equations were usually too complicated to solve. Research on global warming was pushed out to the fringes of science.

Answers started to arrive in the 1950s, and climate change gradually moved back into the mainstream of research. Evidence of rising temperatures had begun to accumulate, and computers could do calculations that were impossible with pencil and paper. The new technique of carbon-14 dating showed how carbon moves between the atmosphere, the oceans, and the biosphere. Careful measurements confirmed that the carbon dioxide content of air was rising. By the '60s, predictions of increasing temperatures again appeared in leading scientific publications.

As the 1970s arrived, the work formerly carried on by a scattering of isolated researchers was developing into a flourishing field of study. Tools were becoming available to sort out the competing effects of carbon dioxide and smoke on climate and calculate how much warmer the earth would get. In the three preceding decades, data showed, there had been a reversal of the trend of rising temperatures. This was the effect of smoke and dust. But the majority of expert opinion held warming to be the likely future trend.

By the end of the decade, science could give an unambiguous answer to the underlying question. In 1979, a committee of the National Academy of Sciences went beyond forecasting and issued an alert. The group offered a consensus prediction: a worldwide temperature increase of 1½ to 4½ degrees if the carbon dioxide in the atmosphere were to double. "A wait-and-see policy may mean waiting until it is too late," its report warned. As when earlier generations of scientists warned of air and water pollution, action would be slow to come.[39]

———

Four years after the death in 1937 of Elon Hooker—he succumbed, as it happened, while vacationing in the Huntington Hotel, the Pasadena hostelry that was to spearhead the cleanup of Southern California's smog[40]—the chemical company he left behind in Niagara Falls needed an outlet for the tarry wastes engendered by burgeoning war production. The eyes of Hooker Electrochemical engineers quickly fell on the old Love Canal. Planned at the end of the nineteenth century to carry the waterpower of the falls to new factories, the canal had been made useless by the invention of alternating electric current that could be transmitted long distances. There remained in 1941 a half mile of clay-bottomed ditch filled with water. The company moved quickly to acquire the land; under the pressure of urgent military demand, the filling of the canal began the next year even before the purchase was complete.

By 1953 the canal was filled with tens of millions of tons of chemical wastes. Hooker covered it with clay and donated the land to the local government for construction of an elementary school. Over the next two decades, a grid of streets was built and a neighborhood of single-family homes grew up around the new school.[41]

Long-standing complaints about chemicals seeping into basements were exacerbated by wet weather in 1976 and 1977, just as scientific studies of pesticide contamination downstream of Niagara Falls began to point toward the city's chemical waste dumps as a source. An investigation by New York State scientists, soon joined by the EPA, rapidly accelerated under pressure from politicians and a local newspaper.

Matters initially came to a head in early August 1978. First the New York State health commissioner, and then President Jimmy Carter, declared a state of emergency. The school was ordered closed, and the state announced that 236 families living nearest the dump would be moved from their homes. A second crisis erupted in May 1980, after a researcher reported genetic damage in more distant portions of the neighborhood. After two EPA officials were demonstratively "taken hostage" by homeowners, another 700 families were evacuated.[42]

The first crisis at Love Canal gained wide publicity in the national media, and other out-of-control hazardous waste sites soon attracted attention. A hazardous waste law, RCRA, had been enacted two years earlier, but so far it was merely a piece of paper—the regulations needed to give it effect had not yet been issued—and it dealt with newly produced wastes, not the legacy of the past. The abandoned dumps were yet another environmental crisis that called for a legislative solution, in a decade when such solutions were sought. When the 1980 evacuation of Love Canal set off an even more intense wave of television and newspaper coverage, pressure became irresistible. The upshot was the creation of the Superfund, approved by a lame-duck Congress in December 1980 and signed into law by the outgoing president on the last day of that year.

The Superfund law made cleanup the responsibility of anyone who had dumped toxic wastes, even if the disposal had occurred decades earlier. The federal government prepared a National Priority List that catalogued the worst sites and set up an enforcement program to ensure that sites on the list were taken care of. When no one could be found to pay for the remedy, the government used a pool of money—the Superfund—collected from a new tax on oil and chemical companies.[43]

The enactment of Superfund brought the study of pollution full circle. Environmental science had always been forward-looking—anticipating harm, looking ahead to technological advance and the new problems and new solutions it would bring. Now it would need to look backwards as well, digging up long-buried dangers, searching for sources, apportioning blame.

CHAPTER 15 | Epilogue: Convenient Hopes and Inconvenient Truths

The potential source of trouble from our change of carbon dioxide in the atmosphere is in the temperature equilibrium.

—Kenneth Banks, 1954[1]

IN THE STRUGGLE OF chemical manufacturers to fend off outside environmental controls, the mid-1950s were a moment of triumph. Congress had passed laws covering air, water, and pesticides, all rejecting national regulation of pollution. The threat of lawsuits had been blunted by safety standards that industry had prescribed for itself. Environmental science was in friendly hands after threatening research programs had been shut down at the Food and Drug Administration, the National Cancer Institute, and the Bureau of Mines. State and local governments, bereft of material and intellectual resources and vulnerable to political and economic pressures, exercised only a feeble oversight.

Why were industry-friendly views so widely adopted among government scientists of the era? This outcome, an early example of a phenomenon now known as "regulatory capture," cannot be explained by the natural process of intellectual interactions among peers. Industry and its allies put forward reasoned arguments, to be sure, but these arguments won more acceptance than their intrinsic persuasive power and the reputations of their proponents can explain. It was often leading scientists and engineers with reputations independent of environmental controversies—Robert Swain, William Mansfield Clark, Morris Fishbein, Harvey Banks, and others—who called attention to problems that were being swept under the rug. Regulatory capture did not happen on its own. It was accomplished by the sustained deployment of political influence and economic inducements, continuing over a period of decades, by the chemical companies and other manufacturing interests.

In the absence of effective outside supervision, the chemical industry understood that responsibility for control of its pollution now fell primarily on its own shoulders. In April 1954, it marked the twin themes of its rapid postwar expansion and its assumption of the burden of environmental protection by convening a Southern Industrial Wastes Conference in Houston. The newly formed Texas Chemical Council and the Southern Association of Science and Industry joined the Manufacturing Chemists' Association as sponsors of this gathering. Some 250 pollution specialists, almost all of them chemical company employees, were in attendance.

The conference opened with a pair of keynote addresses that display the industry's ability to simultaneously hold two contradictory ideas about the genie of pollution it had summoned forth. Alvin Black (1894–1980), chair of the chemistry department at the University of Florida and president of the science and industry association, spoke first. Black reprised the chemical industry's standard talking points, aimed at repelling interference from the outside world. Next followed Kenneth Banks (1914–86), vice president of the New Jersey chemical manufacturer Metal and Thermit Corporation and discoverer of an important anti-leprosy drug. Banks laid out the less sanguine outlook of a scientist who took the long view, explaining to his industry audience what their technological wizardry had unleashed and why it must be brought under control.[2]

Black, a water treatment specialist who ran an engineering business alongside his academic work, opened the conference. His title echoed what the Manufacturing Chemists' Association had proposed two years earlier: "A Rational Approach to the South's Pollution Problem." He offered a message of assurance that there was nothing to fear: "There can be no question that industry recognizes that pollution is bad business.... With the continuance of such an attitude, the South will never face the pollution problem so acute in many parts of the country." A twenty-first-century reader, reaching the end of the paper's seven pages, wonders why Black thought there was a problem at all.

Black's argument was that pollution is the price of progress, and all will turn out for the best if the specialists are left alone to do their work. Like Jack McKee, another water treatment specialist with one foot in academia and the other in consulting, he viewed waste disposal as a legitimate use of waters. This outlook was expressed in words that might have been quoted from Randal Dickey:

> there is no need to prohibit our people and our industries from using streams as an easy and economical means of waste disposal as long as such use does not unreasonably impair their other values.

Water pollution would not be feared, he added, if the public did not expect too much: "Industry cannot be expected to make mountain brooks of creeks ... our air cannot everywhere be clear and clean." Decisions about where the air and water should be clear and clean, and where they should be used as economical means of waste disposal were, of course, to be made locally.

The talk concluded with a précis of the convenient hope and dubious reassurance that the industry offered the public:

> American industry and American governmental units have become aware of these problems and of their individual and collective responsibilities in connection with them. No one can doubt for a moment that the aggressive spirit and the technical "know-how" which have characterized our progress in the past will be more than adequate to solve these problems for the future.[3]

The talk that followed was opposite in tone and content. Where Black parroted industry positions, Kenneth Banks thought for himself. Where Black was firmly fixed in the present, Banks took the long view. Where Black sought to still fears, Banks raised new worries.

Banks, who joked that his job description was Vice President in Charge of the Future, chose as his topic "The Interdependence of Man and Earth." Humanity, he asserted, had changed its relation to nature within the preceding half century through the development of synthetic chemicals and the growing concentration of industry: "Our present industrial and social organization is altering not only the way we live but also the fundamental geological balances of the world in which we live." New stresses on the environment resulted, affecting most of all the supply of fresh water. Taking a broad view, Banks considered how industry was altering global chemical balances. Most chemicals, he asserted, were assimilated by natural processes and caused only

Kenneth Banks teaches a course for advanced high school students about "The Interrelationship of Man and Environment" in 1965. (From the private collection of C. Kenneth Banks, Jr.)

local difficulties, but exceptions caught his eye. Freon from refrigerators is unusually stable, but he thought it would eventually be broken down by sunlight in the atmosphere.

Another worldwide effect of the industrial revolution concerned him much more: "The potential source of trouble from our change of carbon dioxide in the atmosphere is in the temperature equilibrium." A brief survey of warming temperatures followed, from New York to Iceland to Siberia. Whether this was a temporary fluctuation or the onset of a worldwide trend was not certain, but the problem of global warming was real and it demanded attention.

Banks looked back all the way to the dinosaurs, a dominant species that failed to meet a challenge. We do not know, he said, "what challenge we may have to meet. It could be the atomic age—or it could be our tampering with our environment." This was the call to action that environmental specialists in the audience wanted their superiors to hear—when the chair of the MCA air pollution committee, William Burt of B.F. Goodrich Chemical, reported on the Houston meeting to the association's board, he stressed how important it was for top executives to attend pollution conferences.[4]

More than half a century now has passed since that symposium in Houston. The world has come to better recognize the malignant spirits set loose by the chemical industry's wizardry and to search for the formulas that could still them. Yet the rapid forward movement of the '70s has not been sustained, and progress since has been slow and uneven.

One in particular of Kenneth Banks' challenges is still unmet—the inconvenient truth of global warming. The emission of greenhouse gases goes on, protected with the time-honored techniques of toothless laws and twisted science. The tactic of spill, study, and stall, now approaching its centenary, is still in use. Well-funded institutes continue to paste a veneer of scientific research onto political propaganda. Hard truths are countered with convenient but unlikely hopes.

Banks concluded his 1954 talk with a call to his listeners, to the chemical industry, and to the world:

> It is characteristic of our social organization that we strike out boldly into new endeavors and fail to prepare for the reactions such endeavors may produce. We have entered into a socio-industrial revolution that has brought with it basic problems in conservation and waste disposal.... There is only one question: are we going to recognize and put forth the effort to meet the challenge of altered environment? The failure to do so may well be the failure of all for which we have striven... The *greatest* technology we can achieve will do us little good if it causes us to live on a rubbish heap.

Will we meet the challenge of a chemically altered environment? Kenneth Banks' question still awaits its answer.

KEY TO ARCHIVAL CITATIONS

Aut   Unpublished document in possession of authors.

CHF   Chemical Heritage Foundation, Manufacturing Chemists' Association archive.

Dic   Dickey Act legislative history.

Egi   www.egilman.com.

Hag   Hagley Museum. Accession 1801, Box 10, unless otherwise noted.

Hue   Wilhelm Hueper papers, National Library of Medicine.

Ick   Harold Ickes papers, Library of Congress.

ICO   Mutual Chemical files from ICO v. Honeywell, National Library of Medicine.

MCA   Manufacturing Chemists' Association minutes, www.ewg.org.

NCI   Archival collection in director's office, National Cancer Institute.

San   Santa Ana Regional Water Pollution Control Board, file of resolutions.

Tarr   Omar Tarr papers in possession of Douglas Janney.

LIST OF ABBREVIATIONS

CFC	Chlorofluorocarbon
CIO	Congress of Industrial Organizations
CSDS	Chemical Safety Data Sheet
DDT	Dichloro-diphenyl-trichloroethane
EPA	Environmental Protection Agency
FDA	Food and Drug Administration
GM	General Motors
MCA	Manufacturing Chemists' Association
MIT	Massachusetts Institute of Technology
NCI	National Cancer Institute
PCB	Polychlorinated biphenyl
PCE	Perchloroethylene
PHS	Public Health Service
RCRA	Resource Conservation and Recovery Act
TCE	Trichloroethylene

NOTES

Chapter 1

1. "Der Zauberlehrling" (1797), translated by Benjamin Ross.
2. J. L. Meikle, *American Plastic* (Rutgers University Press, 1995), pp. 69–71, has a good overview of this image in the 1920s and 1930s. Quotations are from E. H. Collins, "The Chemical Industry: A Study in Free Enterprise," *New York Times,* Sept. 10, 1951; M. Mears, "Your Nose No Longer Knows," *Washington Post,* Dec. 15, 1929.

Chapter 2

1. D. B. Craig, *After Wilson: The Struggle for the Democratic Party, 1920–1934* (University of North Carolina Press, Chapel Hill, 1992), p. 139.
2. E. Evans, *Trace Metals in Michigan's Ecosystems,* Michigan Dept. of Environmental Quality Report, April 1998, p. 25.
3. P. Brimblecombe, *The Big Smoke: A History of Air Pollution in London since Medieval Times* (Methuen, London, 1987), pp. 9, 15–16.
4. H. Graetz, *History of the Jews* (Jewish Publication Society, Philadelphia, 1894, vol. 4), p. 104.
5. Brimblecombe, *The Big Smoke,* pp. 7–18, 26–36.
6. L. Untermeyer, ed., *Modern British Poetry* (Harcourt, Brace and Howe, New York, 1920).
7. S. H. Dewey, *Don't Breathe the Air* (Texas A&M University Press, College Station, 2000), pp. 15–27; D. Stradling and P. Thorsheim, "The Smoke of Great Cities: British and American Efforts to Control Air Pollution 1860–1914," *Environmental History,* vol. 4, pp. 6–31 (1999).
8. M. L. Quinn, "Industry and Environment in the Appalachian Copper Basin," *Technology and Culture,* vol. 34, pp. 575–612 (1993).

9. D. MacMillan, *Smoke Wars: Anaconda Copper, Montana Air Pollution, and the Courts, 1890–1924* (Helena, Montana Historical Society Press, 2000); F. L. Quivik, "Smoke and Tailings: An Environmental History of Copper Smelting Technologies in Montana, 1880–1930" (PhD diss., University of Montana, 1998); J. E. Lamborn and C. S. Peterson, "The Substance of the Land: Agriculture v. Industry in the Smelter Cases of 1904 and 1906," *Utah Historical Quarterly,* vol. 53, pp. 308–325 (1985); K. G. Aiken, "'Not Long Ago a Smoking Chimney Was a Sign of Prosperity': Corporate and Community Response to Pollution at the Bunker Hill Smelter in Kellogg, Idaho," *Environmental History Review,* vol. 18, pp. 67–86 (1994); J. D. Wirth, *Smelter Smoke in North America* (Lawrence, University Press of Kansas, 2000); R. E. Swain, "Smoke and Fume Investigations, a Historical Review," *Industrial and Engineering Chemistry,* vol. 41, pp. 2384–2388 (1949); J. P. Mitchell, "The Chemical Evidence of Smelter Smoke Injury to Vegetation," *Industrial and Engineering Chemistry,* vol. 8, pp. 175–176 (1916).

10. Plaintiffs' experts who published early scientific reports concerning smelter smoke damage include W. D. Harkins, D. E. Salmon, J. K. Haywood, and G. J. Peirce, as well as Swain: Quivik, "Smoke and Tailings," pp. 328, 331–332; Quinn, "Appalachian Copper Basin," p. 599. The defendant's experts in the Anaconda case also wanted to publish their scientific work, but Anaconda denied them permission to publish: Quivik, pp. 315–323, 332; see also MacMillan, *Smoke Wars,* pp. 110–120. Swain chaired the Stanford University chemistry department from 1917 to 1940 and was acting president of the university from 1928 to 1932: E. F. Hutchison, *The Department of Chemistry, Stanford University, 1891–1976* (Stanford University, Stanford, CA, 1977), pp. 11–13.

11. MacMillan *Smoke Wars,* pp. 152–155, 167–188, 207, quotation from p. 173; Quivik, "Smoke and Tailings," pp. 270–280, 306–344.

12. MacMillan *Smoke Wars,* pp. 185, 231; Quivik, "Smoke and Tailings," pp. 344–372, 419–446; "Montana Smelters Sued," *New York Times,* March 17, 1910.

13. MacMillan *Smoke Wars,* pp. 243–245; Quivik, "Smoke and Tailings," pp. 434–438. The Anaconda smelter produced 14,000 tons of arsenic in 1933 (it is unclear whether this is expressed as As or as As_2O_3): T. LeCain, "The Limits of 'Eco-efficiency': Arsenic Pollution and the Cottrell Electrical Precipitator in the U.S. Copper Smelting Industry," *Environmental History,* vol. 5, pp. 336–351 (2000). Arsenic usage for pesticides in the United States in 1934 can be calculated from P. A. Neal et al., "A Study of the Effect of Lead Arsenate Exposure on Orchardists and Consumers of Sprayed Fruit," Public Health Service Bulletin 267, 1941, p. 12, as approximately 21,000 tons (as As). In this calculation, the average arsenic content of lead arsenate is assumed to be 20% and the annual consumption of Paris green is assumed to be 4.5 million pounds.

14. Dewey, *Don't Breathe the Air,* pp. 30–32; MacMillan, *Smoke Wars,* p. 232.

15. J. D. Wirth, *Smelter Smoke in North America* (University Press of Kansas, Lawrence, 2000).

16. C. E. Colten and P. N. Skinner, *The Road to Love Canal: Managing Industrial Waste Before EPA* (University of Texas Press, Austin, 1996), pp. 69–82; J. I. Leal, "Legal Aspects of Water Pollution," *Public Health,* vol. 27, pp. 103–112 (1901); "The Law of Subterranean Waters," *American Law Register,* vol. 30, pp. 237–264 (1891), esp. pp. 255–260; D. W. Johnson, "Relation of the Law to Underground Waters," U.S. Geological Survey Water-Supply and Irrigation Paper 122, 1903, esp. pp. 25–28; A. S. Travis, "Poisoned Groundwater and Contaminated Soil," *Environmental History,* vol. 2, pp. 343–365 (1997).

17. R. A. Shanley, "Franklin D. Roosevelt and Water Pollution Control Policy," *Presidential Studies Quarterly,* vol. 18, pp. 319–330 (1988); L. B. Dworsky, *Conservation in the United States, A Documentary History: Pollution* (New York, Chelsea House, 1971), p. 21; Office of Technology Assessment, "Preventing Illness and Injury in the Workplace," Report OTA-H-256, April 1985, p. 211; R. C. Williams, *The United States Public Health Service, 1798–1950* (Washington, D.C., Commissioned Officers Association of the United States Public Health Service, 1951), pp. 167, 279–282, 313–320, 535–537.

18. J. A. Tarr, *The Search for the Ultimate Sink* (University of Akron Press, 1996), pp. 354–369; M. M. Cohn and D. F. Metzler, *The Pollution Fighters* (New York State Dept. of Health, 1973), pp. 25–63; W. L. Andreen, "The Evolution of Water Pollution Control in the United States—State, Local, and Federal Efforts, 1789–1972: Part I," *Stanford Environmental Law Journal,* vol. 22, pp. 145–200 (2003).

19. J. Pratt, "The Corps of Engineers and the Oil Pollution Act of 1924," unpublished report prepared for the U.S. Army Corps of Engineers, 1983, pp. 10–24, 43–59; D.C. Drake, "Herbert Hoover, Ecologist: The Politics of Oil Pollution Control, 1921–1926," *Mid-America,* vol. 55, pp. 207–228 (1973), reprinted in R. F. Himmelberg, ed., *Business and Government in America Since 1870* (Garland, New York, 1994, vol. 5), pp. 35–56; H. S. Gorman, *Redefining Efficiency* (University of Akron Press, Akron OH, 2001), pp. 14–20, 23–24, 116. "Inconsequential matters": Craig, *After Wilson,* p. 139.

20. Pratt, "Corps of Engineers," pp. 36–45; Gorman, *Redefining Efficiency,* pp. 20–27.

21. Pratt, "Corps of Engineers," p. 41.

22. Pratt, "Corps of Engineers," pp. 57–68; Gorman, *Redefining Efficiency,* pp. 27–30; F. W. Powell, *The Bureau of Mines* (D. Appleton, New York, 1922), p. 29.

23. Pratt, "Corps of Engineers," pp. 67–78; Drake, "Herbert Hoover, Ecologist"; Gorman, *Redefining Efficiency,* pp. 28, 112–117.

24. Pratt, "Corps of Engineers," pp. 91–105; Gorman, *Redefining Efficiency,* pp. 151–154, 215–224. Hoover vision: D. M. Hart, "Herbert Hoover's Last Laugh: The Enduring Significance of the 'Associative State,'" *Journal of Policy History,* vol. 10, pp. 419–444 (1998); E. W. Hawley, "Herbert Hoover, the

Commerce Secretariat, and the Vision of an 'Associative State,' 1921–1928," *Journal of American History,* vol. 61, pp. 116–140 (1974).

25. Pratt, "Corps of Engineers," pp. 99–100; Gorman, *Redefining Efficiency,* p. 153.

26. Pratt, "Corps of Engineers," pp. 100–110; Shanley, "Roosevelt and Water Pollution"; Dworsky, *Documentary History: Pollution.*

Chapter 3

1. H. B. du Pont, "Management Looks at Air Pollution," in *Proc. Second National Air Pollution Symposium,* Pasadena, May 5–6, 1952, pp. 58–61.

2. A. S. Travis, "Anilines: Historical Background," in Z. Rappoport, ed., *The Chemistry of Functional Groups: The Chemistry of Anilines* (Wiley, Chichester, 2007), pp. 1–73; F. Aftalion, *A History of the International Chemical Industry* (University of Pennsylvania Press, Philadelphia, 1991).

3. P. H. Spitz, *Petrochemicals: The Rise of an Industry* (John Wiley and Sons, New York, 1987), pp. 17–31; Aftalion, *International Chemical Industry,* pp. 35–48.

4. A. D. Chandler, Jr., and S. Salsbury, *Pierre S. Du Pont and the Making of the Modern Corporation* (Harper & Row, New York, 1971), pp. 4–120.

5. Aftalion, *International Chemical Industry,* pp. 34–35, 115–119.

6. D. Whitehead, *The Dow Story* (McGraw-Hill, New York, 1968).

7. D. J. Forrestal, *Faith, Hope & $5,000: The Story of Monsanto* (Simon and Schuster, New York, 1977).

8. W. H. Bower, letter to J. I. Tierney, Feb. 1, 1922; Report of the Executive Committee of the Manufacturing Chemists' Association of the United States, June 18, 1912 (CHF).

9. Aftalion, *International Chemical Industry,* pp. 115–118.

10. E. Russell, *War and Nature: Fighting Humans and Insects with Chemicals from World War I to Silent Spring* (Cambridge University Press, 2001), pp. 17–20; Chandler and Salsbury, *Pierre S. du Pont,* pp. 359–389.

11. Russell, *War and Nature,* pp. 18–21; A. S. Travis, *Dyes Made in America, 1915–1980: The Calco Chemical Company, American Cyanamid, and the Raritan River* (Jerusalem, Hexagon Press, 2004), pp. 37–60; Aftalion, *International Chemical Industry,* pp. 123–124. Government funding of plant construction: Travis, pp. 47, 60; W. Haynes, *American Chemical Industry* (Van Nostrand, New York, 1945), vol. 3, p. 404.

12. Russell, *War and Nature,* pp. 27–32, 41–42; Aftalion, *International Chemical Industry,* p. 124.

13. K. Steen, "Patents, Patriotism, and 'Skilled in the Art': USA v. The Chemical Foundation, Inc., 1923–1926," *Isis,* vol. 92, pp. 91–122 (2001); C. W. Pursell, Jr., "The Farm Chemurgic Council and the United States Department of Agriculture, 1935–1939," *Isis,* vol. 60, pp. 307–317 (1969); Travis, *Dyes Made in America,* p. 69.

14. Travis, *Dyes Made in America,* pp. 50–51, 60–64, 67–79; D. A. Hounshell and J. K. Smith, Jr., *Science and Corporate Strategy: Du Pont R&D, 1902–1980*

(Cambridge University Press, Cambridge, 1988), pp. 88–96; W. S. Dutton, *Du Pont: One Hundred and Forty Years* (Charles Scribner's Sons, New York, 1942), p. 277.

15. A. D. Chandler, Jr., *Strategy and Structure: Chapters in the History of the American Industrial Enterprise* (MIT Press, Cambridge, 1962), pp. 91–113, 124–162; Chandler and Salsbury, *Pierre S. du Pont,* pp. 434–456, 492–536; P. F. Drucker, *Concept of the Corporation* (John Day, New York, 1946).

16. A vast literature in sociology, labor history and other fields discusses the political views of American business. Notwithstanding a great diversity of interpretations, there is wide agreement that the outlook of business elites in the United States has tended toward individualism and has been hostile to outside interference into managers' conduct of their affairs. A useful introduction to these topics from a comparative perspective is S. M. Lipset, *American Exceptionalism* (W.W. Norton, New York, 1997); P. C. Sexton, *The War on Labor and the Left* (Westwood, New York, 1997) gives a historical survey. For a detailed discussion of the du Ponts' views, see R. F. Burk, *The Corporate State and the Broker State: The Du Ponts and American National Politics, 1925–1940* (Harvard University Press, Cambridge, 1990), especially pp. 17–19, 144, 287.

17. Burk, *Corporate State,* pp. 7–8, 25–104; G. Wolfskill, *The Revolt of the Conservatives: A History of the American Liberty League 1934–1940* (Houghton Mifflin, Boston, 1962), pp. 21, 37–43, 46–48, 54–55; Craig, *After Wilson,* pp. 131–156, 181–204; Chandler and Salsbury, *Pierre S. du Pont,* pp. 39, 564, 584–587; P. A. Ndiaye, *Nylon and Bombs* (Johns Hopkins University Press, Baltimore, 2007), pp. 109–116. Craig and Burk document in detail the development of the du Pont group's views as a logical progression from laissez-faire to prohibition to the Democratic party. But it seems unlikely that this was the whole story; a political war against Herbert Hoover was hardly the most natural way to translate laissez-faire philosophy into practical politics. Ndiaye identifies the patent dispute as an influence on the du Ponts' partisan identification, as does T. Ferguson, "From Normalcy to New Deal: Industrial Structure, Party Competition, and American Public Policy in the Great Depression," *International Organization,* vol. 38, pp. 41–94 (1984).

18. Burk, *Corporate State,* pp. 104–191, 246; Wolfskill, *Revolt of the Conservatives,* pp. 20–28, 125–129; Ndiaye, *Nylon and Bombs,* pp. 116–121; Craig, *After Wilson,* pp. 279–283, 285–294.

19. "Elon Hooker Dies, Manufacturer, 68," *New York Times,* May 11, 1938; *Theodore Roosevelt: An Autobiography* (Charles Scribner's Sons, New York, 1913), pp. 284–285; A. Mazur, *A Hazardous Inquiry: The Rashomon Effect at Love Canal* (Harvard University Press, Cambridge, MA, 1998), pp. 9–10; "M.C.A. Presidents—1872–1943," typed list, dated Jan. 6, 1944 (CHF).

20. N. Hapgood, *Professional Patriots* (Albert & Charles Boni, New York, 1927), pp. 51–56, 133, 159–160. Frederic Remington, a director of Great Western

Chemical, which like Hooker specialized in chlorine products, was chairman of the American Defense Society's finance committee.

21. Pursell, "Farm Chemurgic Council"; Travis, *Dyes Made in America,* pp. 93–94; "Almost Joy," *Time,* May 6, 1935; R. P. Browder and T. G. Smith, *Independent: A Biography of Lewis W. Douglas* (Alfred A. Knopf, New York, 1986), pp. 119–131, quotation from p. 124. Douglas, an opponent of tariffs, was not involved in chemical industry lobbying.

22. Wolfskill, *Revolt of the Conservatives,* pp. 144–145, 207, 224–225; Burk, *Corporate State,* pp. 192–235; L. Overacker, "Funds in the Presidential Election of 1936," *American Political Science Review,* vol. 31, pp. 473–498 (1937); Ndiaye, *Nylon and Bombs,* p. 120; Craig, *After Wilson,* pp. 284–285.

23. Burk, *Corporate State,* pp. 236–281; Wolfskill, *Revolt of the Conservatives,* p. 249.

24. Ndiaye, *Nylon and Bombs,* pp. 123–126, 131.

25. E. M. Queeny, *Spirit of Enterprise* (Charles Scribner's Sons, New York, 1943). Queeny board memberships: J. O. Low, review, *American Journal of Sociology,* vol. 49, pp. 376–377 (1944); Russell, *War and Nature,* p. 89. Contributions: Overacker, "Funds in the Presidential Election."

26. Spitz, *Petrochemicals,* pp. 63–69.

27. Spitz, *Petrochemicals,* pp. 220–224; Travis, *Dyes Made in America,* pp. 92–95; J. S. Berman, *The Empire State Building* (Barnes & Noble, New York, 2003), pp. 22–27.

28. Spitz, *Petrochemicals,* pp. 141–146, 153–154.

29. Travis, *Dyes Made in America,* p. 179.

30. Burk, *Corporate State,* pp. 183–189; B. J. Bernstein, "The Truman Administration and the Steel Strike of 1946," *Journal of American History,* vol. 52, pp. 791–803 (1966).

31. Du Pont, "Management Looks at Air Pollution."

Chapter 4

1. Powell, *Bureau of Mines,* pp. 3–6; J. A. Tobey, *The National Government and Public Health* (Johns Hopkins University Press, Baltimore, 1926), p. 296.

2. A. D. Cloud, "Profiles in Occupational Health: Royd Ray Sayers," *Industrial Medicine and Surgery,* vol. 28, pp. 586–591 (1959); http://www.trumanlibrary. org/hstpaper/sayersrr.htm; R. C. Williams, *The United States Public Health Service, 1798–1950* (Washington, Commissioned Officers Association of the United States Public Health Service, 1951), pp. 283, 357–359.

3. A. Derickson, "'Many Have Died and Others Must': The Silicosis Epidemic in Western Hardrock Mining, 1900–1925," in M. L. Hildreth and B. T. Moran, eds., *Disease and Medical Care in the Mountain West* (University of Nevada Press, Reno, 1998), pp. 72–81, 129–34; D. Rosner and G. Markowitz, *Deadly Dust: Silicosis and the Politics of Occupational Disease in Twentieth-Century America*

(Princeton University Press, Princeton, 1991), pp. 31–38. Counted on fingers: Tobey, *National Government and Public Health,* p. 298.

4. "Anthony J. Lanza, N.Y.U. Professor," *New York Times,* March 24, 1964; Rockefeller Foundation, Annual reports, 1921, pp. 92, 136; 1922, p. 87; 1923, pp. 73, 119–120; 1924, p. 75; Cloud, "Royd Ray Sayers"; N. Nelson, "The Earlier Days," copy of a newsletter article (NCI); D. E. Lilienfeld, "The Silence: The Asbestos Industry and Early Occupational Cancer Research—A Case Study," *American Journal of Public Health,* vol. 81, pp. 791–800 (1991).

5. Cloud, "Royd Ray Sayers"; "Dr. Royd Sayers, Noted for Industrial Medicine," *Washington Post,* April 24, 1965.

6. "Former Graphic Publisher Dies in St. Paul," *Lake City (MN) Graphic-Republican,* April 4, 1928; Cloud, "Royd Ray Sayers"; U.S. House of Representatives, *The Investigation of Charges that the Interior Department Permitted the Unlawful Fencing and Inclosure of Certain Lands,* Report 1335 (Government Printing Office, Washington, 1913), quotation from p. 33; "Senator Warren Put Under Fire," *Indianapolis Star,* Jan. 16, 1913; "Would Oust Coach Warner," *New York Times,* May 26, 1914; "Our Indian Wards: Suffer from Ignorance and Politics in Government's Bureau," *New York Times,* Jan. 31, 1915. Linnen's investigations are mentioned in numerous books on Native American history.

7. Cloud, "Royd Ray Sayers"; http://leeboyhoodhome.com/chain.html, accessed August 12, 2008; W. Booth, "Robert E. Lee's Boyhood Home," *Washington Post,* May 13, 1934; J. C. Massey and S. Maxwell, "National Register of Historic Places Inventory—Nomination Form, Robert E. Lee Boyhood Home," Sept. 10, 1985. Society page: "Dr. and Mrs. Sayers to Entertain for Cummings Today," *Washington Post,* Dec. 6, 1936; "Mr. and Mrs. Sayers Will Entertain Indiana People in Alexandria," *Washington Post,* May 20, 1939; "Mrs. R. R. Sayers Will Give Tea Wednesday Afternoon in Honor of Mrs. Paul McNutt," *Washington Post,* Nov. 26, 1939.

8. Loren Kerr, who toward the end of Sayers' career was his subordinate and scientific antagonist, told historian Alan Derickson that Sayers had a pro-industry bias because his wife was from a wealthy mining family. But Edward Linnen lived modestly and was a federal employee for almost his entire career. He did spend six years as a young man publishing newspapers in Lake City and Sibley County, Minnesota, but these are small-town areas far from mining districts. Edna Sayers' mother was from Lake City. There is an 1890 Duluth city directory that lists an Edward B. Linnen as partner in a real estate and investment firm, but it is very doubtful that this business yielded much wealth; Linnen published the Lake City paper at the same time and returned to government service in 1894.

9. The deed is in the Alexandria Public Library. 1799 price: "National Register of Historic Places Inventory—Nomination Form."

10. W. Kovarik, "Charles F. Kettering and the 1921 Discovery of Tetraethyl Lead in the Context of Technological Alternatives," http://www.runet.edu/~wkovarik/

papers/kettering.html, accessed July 12, 2008; W. Kovarik, "The 1920s Environmental Conflict over Leaded Gasoline and Alternative Fuels" (paper presented to American Society for Environmental History, Providence March 26–30, 2003), available at http://www.runet.edu/~wkovarik/papers/ethylconflict.html (accessed Dec. 20, 2009); Hounshell and Smith, *Science and Corporate Strategy*, pp. 127–132; Chandler and Salsbury, *Pierre S. Du Pont*, p. 523. For Midgley's life, see S. B. McGrayne, *Prometheans in the Lab* (New York, McGraw-Hill, 2001), pp. 79–105.

11. G. Markowitz and D. Rosner, *Deceit and Denial* (University of California Press, Berkeley, 2002), p. 18; W. B. Wood, "William Mansfield Clark 1884–1964," *Journal of Bacteriology,* vol. 87, pp. 751–754 (1964).

12. Markowitz and Rosner, *Deceit and Denial,* pp. 18–19; Kovarik, "Context of Technological Alternatives"; C. Warren, *Brush with Death: A Social History of Lead Poisoning* (Johns Hopkins University Press, Baltimore, 2000), pp. 120–123.

13. Markowitz and Rosner, *Deceit and Denial,* pp. 20–21; B. Kovarik, "Agenda Setting in the 1924–1926 Public Health Controversy over Ethyl (Leaded) Gasoline" (paper presented to AEJMC Conference on the Environment, Reno, March 1994); Warren, *Brush with Death,* pp. 117–119, 121–125.

14. Markowitz and Rosner, *Deceit and Denial,* pp. 23–25; Kovarik, "Context of Technological Alternatives"; Warren, *Brush with Death,* pp. 123–126.

15. "Bureau of Mines Put Under Hoover," *New York Times,* June 5, 1925. Lawsuit assistance: L. P. Snyder, "The Death-Dealing Smog over Donora, Pennsylvania: Industrial Air Pollution, Public Health, and Federal Policy, 1915–1963" (PhD diss., University of Pennsylvania, 1994), p. 89. Open publication policy: Powell, *Bureau of Mines,* p. 29. Numerous examples of confidential research by the Bureau of Mines show that the practice was routine between 1925 and 1932. In addition to tetraethyl lead and freon, discussed in this chapter, there are tricresyl phosphate, Hounshell and Smith, *Science and Corporate Strategy,* p. 558; silica dust, D. Rosner and G. Markowitz, "Workers, Industry, and the Control of Information: Silicosis and the Industrial Hygiene Foundation," *Journal of Public Health Policy,* vol. 16, pp. 29–58 (1995); and coal dust, A. Derickson, *Black Lung* (Cornell University Press, Ithaca, 1998), p. 76. Derickson, pp. 71–73, also describes an Industrial Hygiene Division study involving John Bloomfield that relied on the company doctor for medical diagnoses.

16. Markowitz and Rosner, *Deceit and Denial,* pp. 25–34; Kovarik, "Context of Technological Alternatives"; Kovarik, "1920s Environmental Conflict"; M. Neuzil, *Mass Media & Environmental Conflict* (SAGE Publications, Thousand Oaks, CA, 1996), pp. 129–162; Warren, *Brush with Death,* pp. 126–129.

17. Markowitz and Rosner, *Deceit and Denial,* p. 35; D. Davis, *Secret History of the War on Cancer* (Basic Books, New York, 2007), pp. 79, 94–95; J. O. Nriagu, "Clair Patterson and Robert Kehoe's Paradigm of 'Show Me the Data' on Environmental Lead Poisoning," *Environmental Research,* vol. A78, pp. 71–78 (1998); A. P. Loeb, "Paradigms Lost: A Case Study Analysis of Models of Corporate Responsibility for

the Environment," *Business and Economic History,* vol. 28, pp. 95–107 (1999). See also F. Uekotter, "The Merits of the Precautionary Principle," in E. M. DuPuis, *Smoke and Mirrors* (New York University Press, 2004).

18. Snyder, "Death-Dealing Smog," pp. 194–206, 229–245; C. C. Sellers, *Hazards of the Job: From Industrial Disease to Environmental Health Science* (University of North Carolina Press, Chapel Hill, 1997), pp. 160, 172–186, 197–198, 205–207; Warren, *Brush with Death,* pp. 128–132; Davis, *Secret History,* pp. 78–99.

19. Hounshell and Smith, *Science and Corporate Strategy,* pp. 155–157; T. Midgley, Jr., and A. L. Henne, "Organic Fluorides as Refrigerants," *Industrial and Engineering Chemistry,* vol. 22, pp. 542–545 (1930).

20. "Ice Machine Gas Kills 15 in Chicago," *New York Times,* July 2, 1929; "Chicago Restricts Refrigerator Gas," *New York Times,* July 4, 1929; "Refrigerator Inquiry Ordered in Chicago," *New York Times,* July 6, 1929; "Move to Prevent Ice-box Anxiety," *New York Times,* Aug. 1, 1929; "Assure Householders as to Refrigerators," *New York Times,* Aug. 23, 1929. Hounshell and Smith, *Science and Corporate Strategy,* p. 155, confuse the Chicago deaths with a hospital disaster in Cleveland six weeks earlier, which was attributed to nitrogen oxide gases generated by the burning of X-ray films. See "Poison Gas Kills 100 in Cleveland Clinic," *New York Times,* May 16, 1929; "Names Fatal Gases in Cleveland Clinic," *New York Times,* June 16, 1929.

21. Hounshell and Smith, *Science and Corporate Strategy,* pp. 156–157.

22. S. Solomon, "Stratospheric Ozone Depletion: A Review of Concepts and History," *Reviews of Geophysics,* vol. 37, pp. 275–316 (1999).

23. Derickson, *Black Lung,* pp. 69–81. Harrington in Butte: A. Derickson, " Silicosis Epidemic in Western Hardrock Mining,.

24. Cloud, "Royd Ray Sayers." Sources give the date of the transfer variously as 1932 or 1933, but it must have been before Roosevelt's inauguration on March 4, 1933, because it does not appear in a list of reorganizations and transfers after that date printed as an appendix to the 1936 *U.S. Government Manual.* After enactment of the Reorganization Act in July 1932, President Hoover submitted 11 reorganization plans that were all rejected by the Democratic congress; the transfer of the testing laboratory may have escaped congressional scrutiny because the Bureau of Mines' physicians, uniformed Public Health Service officers on detail, were returning to the agency that had always been their nominal employer.

25. Derickson, *Black Lung,* pp. 89–99.

26. M. Cherniack, *The Hawk's Nest Incident* (Yale University Press, New Haven, 1986).

27. Cloud, "Royd Ray Sayers"; Derickson, *Black Lung,* p. 98; A. J. Lanza, memorandum to Dr. Armstrong, Feb. 27, 1935 (Egi); Rosner and Markowitz, *Deadly Dust,* pp. 100–104. Transfer of Bureau of Mines: *U.S. Government Manual,* 1936, p. 3A.

28. Rosner and Markowitz, "Control of Information"; Rosner and Markowitz, *Deadly Dust,* pp. 105–110. Bureau of Mines: "Heads Bureau of Mines," *New York Times,* Aug. 17, 1934; *The Secret Diary of Harold L. Ickes* (Simon & Schuster, New York, 1953), vol. 1, pp. 258, 275; H. L. Ickes, letter to K. T. Compton, Feb. 6, 1940 (Ick).

29. Rosner and Markowitz, "Control of Information"; A. J. Lanza, memorandum to Dr. Armstrong, Feb. 27, 1935 (Egi); V. Brown, letter to C. J. Stover, Dec. 4, 1936 (Egi); Snyder, "Death-Dealing Smog," n. 34, p. 252.

30. Cherniack, *Hawk's Nest Incident,* pp. 74–86; Rosner and Markowitz, "Control of Information."

31. Rosner and Markowitz, "Control of Information"; Cherniack, *Hawk's Nest Incident,* pp. 68–74; Rosner and Markowitz, *Deadly Dust,* pp. 118–120, 132–133; V. Brown, letter to C.J. Stover, Dec. 4, 1936 (Egi); T.C. Waters, "Administration of Laws for the Prevention and Control of Occupational Diseases," *American Journal of Public Health,* vol. 29, pp. 728–737 (1939).

32. Rosner and Markowitz, "Control of Information"; Rosner and Markowitz, *Deadly Dust,* pp. 116–119. H.E. Ayer, "The Origins of Health Standards for Quartz Exposure," *American Journal of Public Health,* vol. 85, pp. 1453–1454, argues that the silica standard was scientifically derived; Rosner and Markowitz reply in the same journal issue.

33. Robert E. Lee Boyhood Home national register nomination; Drew Pearson syndicated column, April 2, 1947.

34. "Dr. Sayers Named to Succeed Mine Director Ickes Ousted," *Washington Post,* April 6, 1940; "Ickes Says Clique Runs Mines Agency," *New York Times,* Jan. 5, 1940; "To Head Bureau of Mines," *New York Times,* April 6, 1940; H. L. Ickes, letter to Civil Service Commissioners, Jan. 31, 1940 (Ick).

35. Derickson, *Black Lung,* p. 110; H. L. Ickes, "Man to Man," *Charleston Gazette,* April 4, 1947.

36. Robert E. Lee Boyhood Home national register nomination; http://leeboyhoodhome.com/chain.html (accessed July 22, 2009).

37. Drew Pearson syndicated column, April 2, 1947.

38. Drew Pearson syndicated column, March 14, 1949; Cloud, "Royd Ray Sayers"; Derickson, *Black Lung,* pp. 115–119, 124–126.

39. Poplar Hill chronology, http://www.poplarhillonhlk.com/timeline.htm, accessed July 6, 2009.

40. Cloud, "Royd Ray Sayers"; "Dr. Royd Sayers, Noted for Industrial Medicine."

Chapter 5

1. E. P. Russell, "The Strange Career of DDT: Experts, Federal Capacity, and Environmentalism in World War II," *Technology and Culture,* vol. 40, pp. 770–796 (1999).

2. J. Whorton, *Before Silent Spring: Pesticides and Public Health in Pre-DDT America* (Princeton University Press, Princeton, 1974), pp. 95–132; A. H. Whitaker,

"A History of Federal Pesticide Regulation in the United States to 1947" (PhD diss., Emory University, 1973), pp. 335–342.

3. Whitaker, "Federal Pesticide Regulation," pp. 342–343; Kelvin et al, *Final Report of the Royal Commission on Arsenical Poisoning* (London, His Majesty's Stationery Office, 1903), ¶¶ 52, 182. Note that arsenic concentrations are stated in the Royal Commission report as As_2O_3; here they are given as As.

4. Whorton, *Before Silent Spring,* pp. 133–142; Whitaker, "Federal Pesticide Regulation," pp. 342–346.

5. Whitaker, "Federal Pesticide Regulation," pp. 346–354, C. O. Jackson, *Food and Drug Legislation in the New Deal* (Princeton University Press, Princeton, 1970), pp. 3–4, 180; Whorton, *Before Silent Spring,* pp. 154–179, 186–200; P. A. Neal et al., "A Study of the Effect of Lead Arsenate Exposure on Orchardists and Consumers of Sprayed Fruit," Public Health Service Bulletin 267, 1941, pp. 2–3.

6. Whitaker, "Federal Pesticide Regulation," pp. 354–360, Jackson, *Food and Drug Legislation,* pp. 89–90, 179–189; Whorton, *Before Silent Spring,* pp. 235–248.

7. Sellers, *Hazards of the Job,* pp. 199–201; Whorton, *Before Silent Spring,* pp. 223–225.

8. House of Commons, "[Preliminary] Report of the Commission on Arsenical Poisoning," *Sessional Papers, Reports from Commissioners, Inspectors, and Others,* 6 July 1901, vol. 9; Kelvin, *Final Report,* ¶¶ 38–41, 169, 182. For more on the history of the precautionary principle, see European Environment Agency, "Late Lessons from Early Warnings: The Precautionary Principle 1896–2000," Environmental Issue Report 22, 2002.

9. Whitaker, "Federal Pesticide Regulation," pp. 360–365; Whorton, *Before Silent Spring,* pp. 225–231; Sellers, *Hazards of the Job,* pp. 209–211; *Treasury Department Appropriation Bill for 1938, Hearings Before the Subcommittee of the Committee on Appropriations, House of Representatives* (Washington, Government Printing Office, 1937), pp. 757–758, 874–876; W. Graebner, "Hegemony through Science: Information Engineering and Lead Toxicology, 1925–1965," in D. Rosner and G. Markowitz, eds., *Dying For Work: Workers Safety and Health in 20th Century America* (Bloomington, Indiana University Press, 1989), p. 150; Markowitz and Rosner, *Deceit and Denial,* pp. 45–62. Neal, "Lead Arsenate Exposure," p. 169, identifies Sayers as the author of the study protocol agreed to between the PHS and Rep. Cannon. Sellers interprets the appropriations clauses as a search for better scientific methodology. But surely a politician like Cannon had little interest in research methodology and wanted to know what results to expect before inserting his rider.

10. Neal, "Lead Arsenate Exposure"; Whorton, *Before Silent Spring,* pp. 244–245.

11. Neal, "Lead Arsenate Exposure," pp. 47, 169; Sellers, *Hazards of the Job,* pp. 219–220. Functions of Horticultural Inspector: Revised Code of Washington, Chapter 15.08.

12. Neal, "Lead Arsenate Exposure," pp. ix, 57; Sellers, *Hazards of the Job,* pp. 211–212.

13. Whitaker, "Federal Pesticide Regulation," pp. 365–373; Whorton, *Before Silent Spring,* pp. 231–233, 244–246; Sellers, *Hazards of the Job,* pp. 209–214; Jackson, *Food and Drug Legislation,* p. 3. Sellers' suggestion that these events were driven by the waxing and waning of scientific methodologies rather than the clash of economic and political interests is as unlikely as his interpretation of the appropriations riders. He describes the crucial July 29, 1940, meeting between the Industrial Hygiene Division and the FDA that decided the new tolerance without noting the presence of McNutt, who was the final arbiter. He suggests several reasons that Neal's point of view might have prevailed, starting with "the PHS's greater prestige," but does not mention the possibility of renewed congressional intervention, which surely weighed on the minds of the participants. Sellers' interpretation is also inconsistent with his own report on p. 219 that subsequent reports of lead poisoning in the population that Neal had studied were suppressed by industry pressure.

14. Whitaker, "Federal Pesticide Regulation," pp. 372–373; Whorton, *Before Silent Spring,* pp. 246–247; Sellers, *Hazards of the Job,* pp. 213–214.

15. Russell, "Strange Career of DDT."

16. Russell, "Strange Career of DDT."

17. E. Russell, *War and Nature: Fighting Humans and Insects with Chemicals from World War I to Silent Spring* (Cambridge University Press, 2001), pp. 113–117, 122–123; Russell, "Strange Career of DDT."

18. Russell, *War and Nature,* pp. 123–127, 147–156; C. Simon, *DDT: Kulturgeschichte einer chemischen Verbindung* (Christoph Merian, Basel, 1999), pp. 43–54, 77–78, 81; Russell, "Strange Career of DDT."

19. Russell, *War and Nature,* pp. 125–126; Russell, "Strange Career of DDT."

20. Russell, *War and Nature,* pp. 127–130, 147–156; Simon, *DDT,* pp. 54–59; Russell, "Strange Career of DDT."

21. Russell, *War and Nature,* pp. 156–159; Russell, "Strange Career of DDT"; Simon, *DDT,* pp. 115–121; "November's Headlines," *Industrial and Engineering Chemistry,* vol. 36, pp. 1177–1178 (1944).

22. Russell, "Strange Career of DDT."

23. Russell, *War and Nature,* pp. 159–160; Russell, "Strange Career of DDT"; Simon, *DDT,* pp. 128–136; "DDT Dangers," *Time,* April 16, 1945.

24. Russell, *War and Nature,* pp. 160–161; Russell, "Strange Career of DDT."

25. Russell, *War and Nature,* p. 163; Russell, "Strange Career of DDT."

26. Whitaker, "Federal Pesticide Regulation," pp. 386–388, 427–428; Russell, *War and Nature,* p. 162.

27. Russell, *War and Nature,* p. 164; Russell, "Strange Career of DDT."

28. Russell, *War and Nature,* pp. 166–171; Russell, "Strange Career of DDT." Typical warnings: "Careful with DDT," *Time,* Oct. 22, 1945; "DDT Spray Called Injurious to Birds," *New York Times,* Oct. 23, 1945.

29. Simon, *DDT,* p. 82; Hounshell and Smith, *Science and Corporate Strategy,* p. 453.
30. Whitaker, "Federal Pesticide Regulation," pp. 413–430.
31. Whitaker, "Federal Pesticide Regulation," pp. 430–434.
32. Whitaker, "Federal Pesticide Regulation," pp. 434–451; C. J. Bosso, *Politics and Pesticides: The Life Cycle of a Public Issue* (University of Pittsburgh Press, Pittsburgh, 1987), pp. 53–58.

Chapter 6

1. H. Heller, "Chromium Dust and Lung Cancer" (memorandum to H. M. Kaufmann, Sept. 1, 1938) (ICO).
2. W. C. Hueper, "Adventures of a Physician in Occupational Cancer: A Medical Cassandra's Tale" (unpublished manuscript), pp. 1–96 (Hue): father's politics, p. 15; volunteering for army, pp. 29–30; 1919 militias, pp. 95–96. R. N. Proctor, *Cancer Wars: How Politics Shapes What We Know and Don't Know About Cancer* (Basic Books, 1995), pp. 36–48, is the best published account of Hueper's life. R. N. Proctor, *The Nazi War on Cancer* (Princeton University Press, 1999), p. 13, says incorrectly of Hueper that "we know from his unpublished autobiography that he had worn the swastika on his Freikorps helmet as early as 1919." For the political evolution of German militias in 1919, see A. Rosenberg, *A History of the German Republic* (Russell & Russell, New York, 1965), pp. 85–99, which is consistent with Hueper's autobiography.
3. Hueper autobiography, pp. 93–141; Hounshell and Smith, *Science and Corporate Strategy,* pp. 558–560. Dates in this portion of Hueper's autobiography are unreliable; see documents cited by Hounshell and Smith and Proctor. German dye manufacturing: A. S. Travis, "Toxicological and Environmental Aspects of Anilines," in Z. Rappoport, ed., *The Chemistry of Functional Groups: The Chemistry of Anilines* (Chichester, Wiley, 2007), pp. 835–870.
4. Hounshell and Smith, *Science and Corporate Strategy,* pp. 561, 563.
5. Hueper autobiography, pp. 142–145; Proctor, *Nazi War on Cancer,* pp. 13–15; Proctor, *Cancer Wars,* pp. 36–38. Proctor in *Nazi War on Cancer* deduces from a letter he located that Hueper signed "Heil Hitler" that Hueper's trip to Germany was motivated in part by an affinity for Nazi ideology. This argument, which is significant because accusations were made after the Second World War that Hueper was a Nazi sympathizer, fails to convince. Proctor's conclusion in his earlier book that "Hueper was by no means a Nazi or an anti-Semite" is the correct one: (1) The letter is not a petition to be allowed to enter Germany, as Proctor says, but a request for a job. (2) Proctor omits the essential fact that when Hueper wrote the letter, he had just been fired from his job and was preparing to go to Germany to look for work. Proctor writes that "It is not always easy to distinguish between conviction and opportunism in such matters." In a job application letter (especially from someone out of work in 1933), it is easy— opportunism can be assumed. (3) Hueper was looking for any work in the area

of pathology, not just in cancer research. Proctor's contention that what German fascists said and did about cancer "might have led a man such as Hueper to bet his future on the Thousand Year Reich" is completely unsupported; Hueper wanted a job in pathology whether or not it had anything to do with cancer. Indeed, Proctor's book documents on pp. 114–119 that in the area of Hueper's research during the 1930s, chemical industry cancers, the Nazis stifled research and prevention activities that had thrived in Germany before 1933. (4) Proctor interprets Hueper's assertion of a desire to restore bonds to German culture as evidence of Nazi sympathies because they were written "only months after the *Machtergreifung*." The introductory words of the sentence—"For a long time"— contradict this. These words are an expression of homesickness, politically neutral but congenial to the intended reader.

6. This story has been told many times: Hueper autobiography, pp. 148–158; Hounshell and Smith, *Science and Corporate Strategy,* pp. 561–564; Proctor, *Cancer Wars,* pp. 38–40; C. Sellers, "Discovering Environmental Cancer: Wilhelm Hueper, Post–World War II Epidemiology, and the Vanishing Clinician's Eye," *American Journal of Public Health,* vol. 87, pp. 1824–1835 (1997); D. Davis, *The Secret History of the War on Cancer* (Basic Books, New York, 2007), pp. 75–77, 91–96; D. Michaels, *Doubt is Their Product* (Oxford University Press, New York, 2008), pp. 21–24.

7. Michaels, *Doubt Is Their Product,* pp. 25–28; Davis, *Secret War on Cancer,* pp. 96–98.

8. H. J. Kaufmann, letter to I.G. Farbenindustrie, Jan. 15, 1936 (ICO); E. Pfeil, "Lungentumoren als Berufserkrankung in Chromatbetrieben," *Deutsche Medizinische Wochenschrift,* vol. 61, p. 1197 (1935).

9. H. M. Kaufmann, memos to H. J. Kaufmann and H. Heller, Aug. 31, 1938; Heller to H. M. Kaufmann, Sept. 1, 1938; H. M. Kaufmann to Heller and H. J. Kaufmann, Sept. 2, 1938 (ICO); "Chrome Dust and Lung Cancer," *Journal of the American Medical Association,* vol. 111, p. 645 (1938). For the editing of *JAMA*'s questions and answers column, see M. Fishbein, *Morris Fishbein, M.D.* (Doubleday, Garden City, NY, 1969), p. 140.

10. Rosner and Markowitz, "Workers, Industry, and the Control of Information"; Rosner and Markowitz, *Deadly Dust,* pp. 115, 117, 128–9; D. E. Lilienfeld, "The Silence: The Asbestos Industry and Early Occupational Cancer Research— A Case Study," *American Journal of Public Health,* vol. 81, pp. 791–800 (1991); Williams, *United States Public Health Service,* pp. 280, 281, 423; W. E. Smith, letter to A. J. Lanza, June 28, 1954 (NCI); J. McCulloch, "Mining and Mendacity, or How to Keep a Toxic Product in the Marketplace," *International Journal of Occupational and Environmental Health,* vol. 11, pp. 398–403 (2005); J. McCulloch and G. Tweedale, *Defending the Indefensible: The Global Asbestos Industry and its Fight for Survival* (Oxford University Press, Oxford, 2008), pp. 70–72; D. L. Egilman and H. H. Hardy, "Corruption of Occupational Medical Literature: The Asbestos Example," *American Journal of Industrial Medicine,*

vol. 20, pp. 127–129 (1991); D. L. Egilman and H. H. Hardy, "Manipulation of Early Animal Research on Asbestos Cancer," *American Journal of Industrial Medicine,* vol. 24, pp. 787–791 (1993); "Anthony J. Lanza, N.Y.U. Professor," *New York Times,* March 24, 1964. Borron et al., "An Early Study of Pulmonary Asbestosis Among Manufacturing Workers: Original Data and Reconstruction of the 1932 Cohort," *American Journal of Industrial Medicine,* vol. 31, pp. 324–334 (1997) [for discussion see vol. 34, pp. 401–410 (1998)] argue, with a focus on the 1930s, that Lanza did not participate in the manufacturers' efforts to prevent publication of research results. Note, however, that Lanza's 1953 actions with respect to Vorwald are reported by Smith in his 1954 letter and a much later interview with David Lilienfeld, and the veracity of Smith's letter is supported by the documentary backing that has now emerged (see below) for its account of Lanza's interactions with Hueper.

11. Proctor, *Cancer Wars,* pp. 40–42; Sellers, "Discovering Environmental Cancer."

12. Hueper autobiography, pp. 166–167; Davis, *Secret War,* pp. 99–100.

13. L. Breslow et al., "Cancer Control: Implications from Its History," *Journal of the National Cancer Institute,* vol. 59, pp. 671–686 (1977).

14. Hueper autobiography, p. 175.

15. W. C. Hueper, "Environmental and Occupational Cancer," *U. S. Public Health Reports,* Supp. 209, 1948; W. C. Hueper, "Environmental Cancer," Federal Security Agency, 1950.

16. Hueper autobiography, pp. 193–202, 223; L. Breslow, ed., *A History of Cancer Control in the United States, 1946–71,* DHEW Publication No. (NIH) 78–1516, 1978; W. C. Hueper, memorandum to J. R. Heller, "Activities of the Section" (memorandum to J. R. Heller, June 8, 1959) (NCI), reprinted in *Drug Research Reports,* September 13, 1961, pp. 355S–360S; W. C. Hueper, "Organized Labor and Occupational Cancer Hazards" (paper presented to AFL-CIO Executive Council, 1959) (Hue); "State Will Study Industrial Cancer," *New York Times,* March 12, 1948.

17. Oscar Ewing, oral history interview, Truman Presidential Library, May 1, 1969, pp. 198–203.

18. Ewing oral history, pp. 203–208; L. P. Snyder, "New York, the Nation, the World: The Career of Surgeon General Thomas J. Parran, Jr.," *Public Health Reports,* vol. 110, pp. 630–632 (1995); "Unhealthy Sequence," *Newsweek,* July 30, 1956, p. 84; "Parran Retired, Scheele Is Named," *New York Times,* Feb. 13, 1948; "Parran Removal, Job Nature Linked," *New York Times,* Feb. 14, 1948; "Leonard Andrew Scheele (1948–1956), Office of the Surgeon General," http://www.surgeongeneral.gov/library/history/bioscheele.htm, accessed June 13, 2008.

19. Davis, *Secret War,* pp. 103–104; Hueper, "Organized Labor"; Hueper, "Activities of the Section." Hueper gives somewhat different accounts of the refusal to reprint his report in his autobiography, p. 249, and in "Activities of the Section."

20. Hueper, "Organized Labor"; Davis, *Secret History,* pp. 102–103; Hueper auto-biography, pp. 176–179.

21. W. H. Hartford, "Fells Point, Baltimore, and Chromium II" (Tarr); Omar F. Tarr obituary, *Lewiston Evening Journal,* May 2, 1955 (Tarr); interviews with Douglas Janney and Jacqueline Tarr Dempsey (Tarr's grandson and daughter-in-law), June 4, 2008.

22. Hartford, Fells Point; G. A. Benington, memo to O. F. Tarr, Dec. 27, 1945 (ICO). Death from lung cancer: W. M. Gafafer, ed., *Health of Workers in Chromate Producing Industry,* U.S. Public Health Service Publication 192, 1953, p. 4.

23. Hueper autobiography, p. 176; Hueper, "Organized Labor," pp. 7, 8; Gafafer, *Chromate Producing Industry,* p. 4.

24. Gafafer, *Chromate Producing Industry,* p. 4. Tarr is identified as the traveler to Germany in his *Lewiston Evening News* obituary.

25. Gafafer, *Chromate Producing Industry,* p. 4, states that the chromium industry's study of lung cancer was triggered by a 1945 death that was "alleged" to have been caused by lung cancer. The date of the allegation is not stated. The location of the death or deaths in Jersey City is shown by Omar Tarr's practice of filing correspondence about lung cancer under "Jersey City compensation." There may have been multiple claims; Tarr's Jan. 30, 1947 memo to J. Dohan (ICO) refers to "compensation cases" in the plural.

26. It is not clear whether Waters' Baltimore law firm was hired for this purpose or already represented Mutual; in 1948 the firm handled the real estate trans-actions associated with the company's new plant.

27. T. C. Waters, letter to A. J. Lanza, Feb. 3, 1947 (ICO). There is other cor-respondence in which the problem is unnamed: Benington to Dohan, Feb. 3, 1947; Benington to Tarr, Jan. 28, 1947; Waters to A. T. Vanderbilt, Jan. 28, 1947 (ICO).

28. W. Machle and F. Gregorius, *Cancer of the Respiratory System in the U.S. Chro-mate-producing Industry,* Public Health Reports, vol. 63, pp. 1114–1127, acknowledge Lanza's assistance. The sequence of events is established by comparing (1) Waters' letter to Lanza, Feb. 3, 1947 (ICO), which mentions a meeting the previous Saturday and says "We will proceed in the matter in accordance with your recommendations," with (2) the Machle and Gregorius paper, which states that in 1947 "one of the large producers of chromates in the United States became concerned with the incidence of lung cancer among their employees" and then describes a single-company analysis followed by the industry-wide study. Machle formerly at Kettering: W. Machle, F. F. Heyroth, and S. Witherup, "The Fate of Methylcellulose in the Human Digestive Tract," *Journal of Biological Chemistry,* vol. 153, pp. 551–559 (1944); W. Machle and E. W. Scott, "The Effects of the Inhalation of Hydrogen Fluoride. III. Fluorine Storage Following Exposure to Sub-Lethal Concentration," *Journal of Indus-trial Hygiene,* vol. 17, pp. 230–240 (1935). For more on Machle, see Michaels, *Doubt Is Their Product,* pp. 125, 127.

29. Machle and Gregorius, who do not mention Mutual's consultation with Hueper, date the beginning of the company's concern about lung cancer to 1947, while Gafafer dates it to 1945; the memoranda cited above show that it actually began no later than the 1930s.

30. Machle and Gregorius, *Cancer of the Respiratory System in the U.S. Chromate-producing Industry.*

31. O. F. Tarr, letter to A. E. Van Wirt, Imperial Paper & Color Corp., Dec. 27, 1948; O. F. Tarr, letter to C. D. Marlatt, Martin Dennis Company, Dec. 30, 1948; A. J. Dohan, letter to S. Cohen, Dec. 30, 1950 (ICO).

32. Hartford, Fells Point; L. C. Palmer and G. E. Best, "Air Pollution Control in a Chromium Chemicals Plant," *Industrial Hygiene Quarterly,* pp. 294–298 (1953); J. A. Benington, letter to Mrs. Tarr, April 19, 1949 (Tarr); "Chemical Plant Is World's Greatest in Particular Line," *Baltimore American,* April 2, 1951 (Tarr). Best's hiring: Tarr to Benington, Feb. 26, 1948 (ICO). DuPont's environmental investment was $12.6 million compared to a total plant value of $1.4 billion: G. M. Read, "Summary of Figure Data—Industrial Department Reports on Water and Air Pollution (memo to DuPont Executive Committee, Aug. 9, 1950) (Hag), and Hounshell and Smith, *Science and Corporate Strategy,* p. 603. Hartford's recollection, in Fells Point and in a Mutual Chemical chronology he prepared in 1992 (CHF, Winslow Hartford collection), was that the need for a new plant was understood before the war and work began immediately after it. It is possible that this is mistaken; Benington's 1945 memo seems to envisage only improvements to the existing plant, and July 7 and July 9, 1948 memos from Tarr to J. Dohan and A. G. Noble and to Heller, L. C. Palmer, and H. B. Smith (ICO) suggest that the design of the new plant was just beginning. Hartford worked under Tarr and, writing years afterward, might have confused his boss's opinion with company policy. Another possible explanation is that Mutual was concerned about undermining its defense against liability suits and therefore delayed starting work on the new plant until it had completed its charade of discovering the lung cancer problem.

33. T. F. Mancuso and W. C. Hueper, "Occupational Cancer and Other Health Hazards in a Chromate Plant: A Medical Appraisal, I. Lung Cancer in Chromate Workers," *Industrial Medicine & Surgery,* vol. 20, pp. 358–363 (1951).

34. *Food Additives, Hearings Before a Subcommittee of the Committee on Interstate and Foreign Commerce, House of Representatives,* Government Printing Office, 1958: Testimony of W. E. Smith, July 18–19, 1957, pp. 172, 182–185; W. C. Hueper, letter to W. E. Smith, Aug. 24, 1955, ibid., pp. 188–189; W. E. Smith, letter to A. J. Lanza, June 28, 1954 (NCI); O. F. Tarr, letter to A. M. Baetjer, Oct. 30, 1953 (ICO); Hueper, "Organized Labor," p. 8; Hueper autobiography, pp. 176, 179, 250; Hueper, "Activities of the Section," pp. 11–12 (NCI); M. B. Shimkin, "As Memory Serves—An Informal History of the National Cancer Institute, 1937–57," *Journal of the National Cancer Institute,* vol. 59, pp. 559–600 (1977). Lanza's denial of misconduct is on p. 186 of the hearings; a denial

from Surgeon General Leroy Burney—which admits that the scope of Hueper's activities had been curtailed but asserts purely administrative motives—is on pp. 190–194. The denials are contradicted by Shimkin, a memoirist not friendly to Hueper, and Lanza's involvement is corroborated by the Tarr letter to Baetjer, which states that "Mancuso's paper with this statement came to Doctor Lanza's attention before it was published and Doctor Lanza wrote Mancuso that in his opinion the implications related to this particular incident were quite unjustified and therefore should be omitted from the paper."

35. Hueper to Smith, Aug. 24, 1955.

36. Tarr to Hueper, June 29, 1953; Hueper to Tarr, Oct. 30, 1953; Hueper to Tarr, Nov. 30, 1953; Tarr to Benington, Dec. 1, 1953; Tarr to Hueper, Dec. 1, 1953; Tarr to Hueper, Dec. 22, 1953; Hueper to Smith, Aug. 24, 1955; Hueper, "Activities of the Section."

37. "Occupational Health's Dynamo," *Johns Hopkins Public Health Magazine,* Fall 2001.

38. Hartford, Fells Point.

39. O. F. Tarr, letter to P. W. Dilthey, July 22, 1953; O. F. Tarr, memorandum to file, Dec. 15, 1953 (ICO). See also Tarr's memorandum to file, Dec. 9, 1953 (ICO).

40. J. Tarr Dempsey interview. Anna Baetjer, in a condolence letter (Tarr), wrote that "This is one of the most tragic things that could have happened to the man who has done so much to prevent illness of this type in his plant."

Chapter 7

1. M. Brienes, "The Fight against Smog in Los Angeles, 1943–1957" (PhD diss., University of California at Davis, 1975), p. 120.

2. The coinage is commonly attributed to des Voeux, as reported in a July 26, 1905, newspaper account—variously described as appearing in either the *Daily Gazette* or *Daily News*—of a talk he gave. But the word appeared earlier, in an untitled note about London in the January 19, 1893, *Los Angeles Times,* and the newspaper article about des Voeux's talk refers to "what was known as" smog. See also G. A. Bergström, "On Blendings of Synonymous or Cognate Expressions in English," PhD diss., University of Lund, 1906, p. 61.

3. A. Novakov and T. Novakov, "Eyewitness: The Chromatic Effects of Late Nineteenth-century London Fog," *Literary London,* vol. 4, http://www.literary-london.org/london-journal/september2006/novakov.html (2006).

4. B. Herzog, "Louis C. McCabe," Illinois State Geological Survey citation, http://www.isgs.uiuc.edu/about-isgs/heritage/mccabe.shtml, accessed Aug. 18, 2008; D. Stradling, *Smokestacks and Progressives* (Johns Hopkins University Press, Baltimore, 1999), p. 166.

5. Stradling, *Smokestacks and Progressives* pp. 163–175; Brienes, "Fight against Smog," pp. 17–22; J. A. Tarr and B. C. Lamperes, "Changing Fuel Use Behavior and Energy Transitions: The Pittsburgh Smoke Control Movement,

1940–1950," *Journal of Social History,* vol. 4, pp. 561–588 (1981); C. O. Jones, *Clean Air: The Policies and Politics of Pollution Control* (University of Pittsburgh Press, Pittsburgh, 1975), pp. 23–29, 43–47.

6. S. H. Dewey, *Don't Breathe the Air: Air Pollution and U.S. Environmental Politics, 1945–1970* (Texas A&M University Press, 2000), chap. 3; Brienes, "Fight against Smog," pp. 1–11. See also J. Dunsby, "Localizing Smog: Transgressions in the Therapeutic Landscape," in E. M. DuPuis, ed., *Smoke and Mirrors* (New York University Press, 2004), pp. 170–200.

7. Brienes, "Fight against Smog," pp. 32–53, also published as "Smog Comes to Los Angeles," *Southern California Quarterly,* vol. 58, pp. 515–32 (1976). "Inevitable": Brienes, "Fight against Smog," p. 85.

8. Brienes, "Fight against Smog," pp. 58–90. Use of "smog": p. 29, n. 13.

9. Brienes, "Fight against Smog," pp. 92–95.

10. Brienes, "Fight against Smog," pp. 95–99, 155–164.

11. Brienes, "Fight against Smog," pp. 99–107, 147–150.

12. Brienes, "Fight against Smog," pp. 101–102, 107–108, 168, n. 30; J. Hulse, "Strange City of Vernon Booms by Day, Dies Each Night," *Los Angeles Times,* Nov. 22, 1953.

13. Brienes, "Fight against Smog," pp. 115–130; Dewey, *Don't Breathe the Air,* pp. 42–43; "The Showdown on Smog," *Los Angeles Times,* May 18, 1947; "The New World," *Time,* July 15, 1957.

14. Herzog, Louis C. McCabe.

15. Brienes, "Fight against Smog," pp. 131–137, 164, 169–176; C. L. Senn, "Los Angeles 'Smog,'" *American Journal of Public Health,* vol. 38, pp. 962–965 (1948); Dewey, *Don't Breathe the Air,* pp. 44–47; A. J. Haagen-Smit, "A Lesson from the Smog Capital of the World," *Proceedings of the National Academy of Sciences,* vol. 67, pp. 887–897 (1970).

16. Brienes, "Fight against Smog," p. 217.

17. Brienes, "Fight against Smog," pp. 181–186; Dewey, *Don't Breathe the Air,* p. 48.

18. Brienes, "Fight against Smog," pp. 186–203; Dewey, *Don't Breathe the Air,* pp. 48–50; A. J. Haagen-Smit, "Formation of Ozone in Los Angeles Smog," *Proc. Second National Air Pollution Symposium,* Pasadena, May 5–6, 1952, pp. 54–56.

Chapter 8

1. "Public Health Service Plans Survey in Pennsylvania—40 Donora Sufferers Fly South," *New York Times,* Nov. 19, 1948; J. G. Townsend, "Investigation of the Smog Incident in Donora, Pa., and Vicinity," *American Journal of Public Health,* vol. 40, pp. 183–189 (1950).

2. H. H. Schrenk, H. Heimann, G. D. Clayton, and W. M. Gafafer, *Air Pollution in Donora, Pa.: Epidemiology of the Unusual Smog Episode of October 1948. Preliminary Report,* Public Health Bulletin No. 306, Division of Industrial Hygiene,

1949, pp. 86–103; Snyder, "Death-Dealing Smog," p. 35. Part of chemical industry: Zinc is separated from ore as a vapor and condensed, while other smelting processes collect liquid metal. Much of it was sold as a powdered oxide chemical, with paint companies as major customers, rather than as a metal. A zinc plant upriver from Donora belonged to DuPont.

3. Snyder, "Death-Dealing Smog," pp. 22–28; D. Davis, *When Smoke Ran Like Water* (Basic Books, 2002), pp. 5–18; B. Roueché, *Eleven Blue Men* (Little Brown, Boston, 1953), pp. 194–211.

4. Snyder, "Death-Dealing Smog," pp. 22–23; M. Firket, "Sur les causes des accidents survenus dans la vallée de la Meuse, lors des brouillards de décembre 1930," *Bulletin de l'Académie Royale de Médecine de Belgique*, vol. 11, pp. 683–732 (1931). For the initial blame in the Meuse: "Scores Die, 300 Stricken by Poison Fog in Belgium; Panic Grips Countryside," *New York Times*, Dec. 6, 1930; "Fog Brought Death Only to Old and Ill," *New York Times*, Dec. 7, 1930; W. P. Carney, "Queen Backs Work in Poison Fog Area," *New York Times*, Dec. 14, 1930. In Donora, Lynn Snyder, "Death-Dealing Smog," pp. 30–31, 67, reports that "Investigators from outside the Donora district were uniformly struck by the confidence of area residents that the Zinc Works had caused the smog disaster." See also "Donora Asks U.S. Inquiry," *New York Times*, Nov. 3, 1948; Davis, *Smoke Like Water*, pp. 17–26.

5. Snyder, "Death-Dealing Smog," pp. 27–30. Nearly half: Schrenk, *Air Pollution in Donora*, p.163.

6. Snyder, "Death-Dealing Smog," pp. 29–31; "Death Smog Toll Hits 20 as Hundreds Flee Town," *Los Angeles Times*, Nov. 1, 1948; "20 Dead in Smog; Rain Clearing Air as Many Quit Area," *New York Times*, Oct. 31, 1948; "Donora Asks U.S. Inquiry." Davis, *When Smoke Ran Like Water*, pp. 27–28, confirmed that the deaths were centered around the zinc mill. Los Angeles regulations: Brienes, "Fight against Smog," p. 169; G. P. Larson, "Reduction at the Source," *Proc. First National Air Pollution Symposium*, Pasadena, Nov. 10–11, 1949, pp. 77–79.

7. Snyder, "Death-Dealing Smog," p. 28–35; "Donora Asks U.S. Inquiry."

8. Snyder, "Death-Dealing Smog," p. 38; Roueché, *Eleven Blue Men*, p. 215.

9. Election results from Washington County Election Office. Progressive Party candidate Henry Wallace received 81 votes in Donora.

10. Snyder, "Death-Dealing Smog," pp. 39–44; p. 150, n. 11; p. 180, n. 6. Snyder suggests in her text, p. 148, that the involvement of the Industrial Hygiene Division was triggered by a recommendation from Wesley Hemeon, the Industrial Hygiene Foundation investigator, to U.S. Steel. This interpretation is contradicted by the sources in Snyder's footnote and several additional pieces of evidence. It cannot explain the Industrial Hygiene Division's simultaneous reversal on its chromium study, which as discussed below was unquestionably contrary to the wishes of the industry. In the context of the national political situation, a sudden post-election policy reversal would much more likely

respond to labor than management. Internal Steelworkers correspondence, J. Richardson, letter to F. Burke, Nov. 18, 1949, Philip Murray Papers, Box 62, File 14, Catholic University of America, ascribes the PHS involvement to "your efforts and those of others in the organization." The interpretation in Snyder's footnote is the correct one; the IHF and the company would have preferred their own investigations to those by the government.

11. Schrenk et al., *Air Pollution in Donora,* in an introductory discussion of the study's origin, says that "On Tuesday morning a telephone call came to the Division of Industrial Hygiene..." From where the call came is not stated, but we are told that the request "was later formally repeated on behalf of the Borough Council of Donora, the Department of Health of the State of Pennsylvania, and the United Steelworkers of America, CIO."

12. "The President's Day, Friday, November 5, 1948," Truman Presidential Library, http://www.trumanlibrary.org/calendar/main.php?currYear=1948&currMonth=11&currDay=5. The calendar does not show whether Ewing was still in the room when Murray arrived.

13. Richardson, letter to Burke, Nov. 18, 1948.

14. Snyder, "Death-Dealing Smog," pp. 144–150; "Federal Experts Will Study Smog," *New York Times,* Nov. 19, 1948; O. F. Tarr, Mutual Chemical Company, letter to C. D. Marlatt, Martin Dennis Company, Dec. 30, 1948 (ICO).

15. "Denies Smog Zinc Blame," *New York Times,* Nov. 17, 1948; Snyder, "Death-Dealing Smog," pp. 207–214.

16. Tarr, letter to Marlatt, Dec. 30, 1948.

17. Snyder, "Death-Dealing Smog," pp. 149–173, 215–222. Snyder, p. 61, n. 73, reports that, except for a single memorandum, no records of the investigation seem to have been preserved in government archives. Schrenk collaboration with Sayers: see, e.g., W. P. Yant, H. H. Schrenk, and R. R. Sayers, "Methanol Antifreeze and Methanol Poisoning," *Industrial and Engineering Chemistry,* vol. 23, pp, 551–555 (1931). For another conflict over control of histological work, see W. C. Hueper, memorandum to G. Seger, May 23, 1952 (NCI).

18. Snyder, "Death-Dealing Smog," pp. 35–36, 152–155; "Donora Smog Held Near Catastrophe," *New York Times,* Dec. 25, 1948. Inadequacy of funds: Richardson to Burke, Nov. 18, 1948.

19. Snyder, "Death-Dealing Smog," pp. 33, 164–168, 178–179.

20. Schrenk et al., *Air Pollution in Donora, Pa.;* Townsend, "Investigation of the Smog Incident"; B. Furman, "Government Spurs Poisoned Air Study," *New York Times,* Oct. 14, 1949.

21. Furman, "Government Spurs Poisoned Air Study"; C. A. Mills, "The Donora Episode," *Science,* vol. 111, pp. 67–68 (1950).

22. Snyder, "Death-Dealing Smog," pp. 91, 210–229.

23. Snyder, "Death-Dealing Smog," p. 261, n. 85.

24. A. J. Lanza, "Health Aspects of Air Pollution" (paper presented to Smoke Prevention Association of America, Birmingham, May 24, 1949).

25. Dewey, *Don't Breathe the Air,* p. 45; L. C. McCabe, "National Trends in Air Pollution," *Proc. First National Air Pollution Symposium,* Pasadena, Nov. 10–11, 1949; G. Hill, "Lag on Smog Laid to Industrialists," *New York Times,* Nov. 11, 1949; excerpts also appeared in the *Los Angeles Times* and *Chicago Tribune.*

Chapter 9

1. M. Trumper, "Water Pollution," in Senate Committee on Commerce, *Stream Pollution, Hearings before a Subcommittee of the Committee on Commerce, United States Senate* (Government Printing Office, Washington, 1936), pp. 403–404.

2. Shanley, "Roosevelt and Water Pollution"; Dworsky, *Documentary History: Pollution;* Williams, *The United States Public Health Service,* pp. 167, 535–538.

3. P. V. Scarpino, *Great River: An Environmental History of the Upper Mississippi, 1890–1950* (University of Missouri Press, Columbia, 1985), pp. 114–150.

4. *Stream Pollution Hearings,* p. 55; T. Kehoe, *Cleaning Up the Great Lakes: From Cooperation to Confrontation* (Northern Illinois University Press, Dekalb, 1997), pp. 23–24, 29–32, 35; Andreen, "Water Pollution Control: Part I," pp. 186–188.

5. Scarpino, *Great River,* pp. 156–158; W. L. Andreen, "The Evolution of Water Pollution Control in the United States—State, Local, and Federal Efforts, 1789–1972: Part II," *Stanford Environmental Law Journal,* vol. 22, pp. 215–294 (2003); Shanley, "Roosevelt and Water Pollution"; *Stream Pollution Hearings.*

6. Quoted in Pratt, "Corps of Engineers," p. 93.

7. J. K. Smith, "Turning Silk Purses into Sows' Ears: Environmental History and the Chemical Industry," *Enterprise and Society,* vol. 1, pp. 785–812 (2000); Pratt, "Corps of Engineers," pp. 92–93.

8. *Stream Pollution Hearings,* pp. 157–165, 180, quotation from p. 161; W. Hollander, Jr., *Abel Wolman: His Life and Philosophy, an Oral History* (Universal Printing & Publishing, Chapel Hill, NC, 1981), vol. 2, pp. 1085–1086.

9. Hollander, *Abel Wolman,* pp. 591–596, 699–700.

10. M. Reutter, *Making Steel: Sparrows Point and the Rise and Ruin of American Industrial Might* (University of Illinois Press, Urbana, 2004), pp. 336–341, 399–404. Political connections: Hollander, *Abel Wolman,* pp. 707–712, 716.

11. "CS2 Poisoning," *Time,* March 18, 1940; P. J. Kuznick, *Beyond the Laboratory: Scientists As Political Activists in 1930s America* (University of Chicago Press, 1987), p. 239.

12. *Stream Pollution Hearings,* pp. 402–404.

13. *Stream Pollution Hearings,* pp. 228–231; Smith, "Silk Purses into Sows' Ears."

14. Smith, "Silk Purses into Sows' Ears."

15. Smith, "Silk Purses into Sows' Ears."

16. K. A. Reid, letter to *New York Times,* July 20, 1939; W. F. Wiley, letter to *New York Times,* July 27, 1939; Shanley, "Roosevelt and Water Pollution"; Pratt, "Corps of Engineers," pp. 109–111; R. R. Camp, "Wood, Field, and Stream," *New York Times,* Feb. 28, 1940. Tarr, *Search for the Ultimate Sink,* pp. 370–373, ascribes to industry only a minor role in the failure of New

Deal efforts to regulate water pollution. Tarr, Christopher Sellers, and some other historians of pollution control have focused on professional culture as a determinant of opinions about environmental regulation. But a systematic correlation between professional affiliations and opinions about regulation is hard to find. In post–Second World War air pollution debates, engineers favored greater control than public health specialists, while in early twentieth-century water pollution debates, public health specialists favored greater control than engineers. The Agriculture Department's entomologists and the Public Health Service swapped positions on DDT between the mid-1940s and the 1960s.

17. Andreen, "Water Pollution Control: Part II"; Shanley, "Roosevelt and Water Pollution"; Dworsky, *Documentary History: Pollution;* J. C. Davies III, *The Politics of Pollution* (Pegasus, New York, 1970), pp. 38–40; Pratt, "Corps of Engineers," pp. 110–115.

18. Colten and Skinner, *Road to Love Canal,* pp. 138–142; Pratt, "Corps of Engineers," pp. 115–116.

19. Andreen, "Water Pollution Control: Part I."

20. Scarpino, *Great River,* pp. 160–161; W. Murphy, "Industrial Wastes . . . A Chemical Engineering Approach to a National Problem," *Industrial and Engineering Chemistry,* vol. 39, pp. 557–558 (1947); K. B. Brooks, *Before Earth Day: The Origins of American Environmental Law, 1945–1970* (University Press of Kansas, Lawrence, 2009), pp. 28–29, 53.

21. W. B. Hart, "Disposal of Petroleum Refinery Wastes, Article 1—Industry's Relation to the Problem," *National Petroleum News,* vol. 38, pp. R-11—R-16 (1946).

22. Murphy, "Industrial Wastes"; A. Wolman, "State Responsibility in Stream Pollution Abatement," *Industrial and Engineering Chemistry,* vol. 39, pp. 561–565 (1947); S. T. Powell, "Creation and Correction of Industrial Wastes," *ibid.,* pp. 565–568.

23. T. Parran, "The Public Health Service and Industrial Pollution," *Industrial and Engineering Chemistry,* vol. 39, pp. 560–561 (1947).

24. Andreen, "Water Pollution Control: Part II"; P. C. Milazzo, *Unlikely Environmentalists: Congress and Clean Water, 1945–1972* (University Press of Kansas, Lawrence, 2006), pp. 19–20; Dworsky, *Documentary History: Pollution,* p. 25; Smith, "Silk Purses into Sows' Ears." Section 2(d) of the law allowed the Public Health Service to seek injunctions against discharges in one state that affected public health or welfare in another state, but only if permission was granted by the regulatory agency of the state in which the pollution originated.

25. Scarpino, *Great River,* p. 162.

Chapter 10

1. C. G. Hanson, "Water Pollution Action Postponed," *Los Angeles Times,* June 8, 1949.

2. Colten and Skinner, *Road to Love Canal,* pp. 31–39.

3. M. Deutsch, *Ground-water Contamination and Legal Controls in Michigan,* U.S. Geological Survey Water-Supply Paper 1691 (Washington, 1963); N. M. Perlmutter, M. Lieber, and H. L. Frauenthal, "Movement of Waterborne Cadmium and Hexavalent Chromium Wastes in South Farmingdale, Nassau County, Long Island, New York," in *Short Papers in Geology and Hydrology,* Articles 60–121, U.S. Geological Survey Professional Paper 475-C, 1963, pp. C179–C184; D. Graham, "Chromium: A Water and Sewage Problem," *Journal of the American Water Works Association,* vol. 35, p. 159 (1943); R. R. Bennett and R. Meyer, "Geology and Ground-water Resources of the Baltimore Area," *Maryland Geological Survey Bulletin* 4, 1952, p. 132.

4. Deutsch, *Ground-water Contamination,* p. 30; D. J. Cederstrom, "The Arlington Gasoline-Contamination Problem" (unpublished U.S. Geological Survey report, 1947), abstract in *Bibliography of Publications Relating to Ground Water Prepared by the Geological Survey and Cooperating Agencies 1946–55,* U.S. Geological Survey Water-Supply Paper 1492, 1957, p. 28. No date is given for the study mentioned in Deutsch's 1963 report, but the majority of pollution incidents described in that report occurred in the 1940s.

5. H. A. Swenson, "The Montebello Incident," *Proc. Soc. Water Treatment Examination,* vol. 11, pp. 84–88 (1962); A. Pickett, "Protection of Underground Water from Sewage and Industrial Wastes," *Sewage Works Journal,* vol. 19, pp. 464–472 (1947); Engineering-Science Inc., *Effects of Refuse Dumps on Ground Water Quality,* California State Water Pollution Control Board Publication 24, 1961, pp. 91–92; D. B. Willets and M. L. Gould, "Ground-water—A Vulnerable Resource," *Proceedings of the International Association of Hydrological Sciences,* Berkeley, Aug. 13–31, 1963, IAHS Pub. 63, pp. 482–492. Descriptions of the material discharged from the plant differ somewhat; this account is based on Swenson. Swenson says that the incident was chronicled in the press, but we were not able to find any reference in the *Los Angeles Times* archives.

6. N. Quam-Wickham, " 'Cities Sacrificed on the Altar of Oil': Popular Opposition to Oil Development in 1920s Los Angeles," *Environmental History,* vol. 3, pp. 189–209 (1998); Allen W. Hatheway, "Pre-RCRA History of Industrial Waste Management in Southern California," in B. W. Pipkin and R. J. Proctor, eds., *Engineering Geology Practice in Southern California,* Association of Engineering Geologists, Southern California Section, Special Publication No. 4 (Belmont, Cal., 1992), pp. 35–65.

7. H. A. Thomas and D. A. Phoenix, *Summary Appraisals of the Nation's Ground-Water Resources—California Region,* U. S. Geological Survey Professional Paper 813-E, 1976.

8. The prohibitions had been recodified as California Health and Safety Code, §§5410, 5417. §5417 banned unpermitted discharge of "sewage," which was defined in §5410 to include "Any animal, mineral or vegetable matter or

substance offensive, injurious or dangerous to health." This definition clearly included toxic industrial wastes. §5422 required the permit application to describe the treatment plant. §5432 provided that the manager of the treatment plant, although paid by the permittee, must be approved by the Health Department. Penalties were imposed by §§5463–64.

9. *Report of the Interim Fact-Finding Committee on Water Pollution,* Assembly of the State of California, 1949, referred to below as "Dickey Report," p. 74.

10. See, e. g., C. H. Purcell, Director of Public Works, memorandum to Gov. Earl Warren, June 21, 1949, p. 13 (Dic).

11. Dickey Report, p. 168. Chemical industry participation: *Chemical and Engineering News,* vol. 26, p. 2822 (Sept. 20, 1948).

12. Dickey Report, quotation from p. 7. In the terminology of the time, "mineral wastes" included organic chemicals not derived from living organisms; for example, the report on p. 74 refers to "organic minerals such as phenols."

13. Dickey Report, p. 69.

14. The second paragraph of the report refers to "outbreaks of intestinal disease from sewage-polluted public water supplies" [p. 7]; a typhoid epidemic is reported to have occurred while the study was in progress [p. 9]; and the first conclusion stated at the end is that "Public health is not protected from the effects of sewage-polluted waters…the result of improperly controlled sewage disposal facilities" [p. 105]. On the other hand, when the danger of ground-water pollution by industrial wastes and the tendency of such pollution to be long-lasting are admitted, it is quickly asserted that such pollution is rarely "widespread" [p. 15].

15. Dickey Report, pp. 71, 172.

16. Dickey Report, p. 105, emphasis in original. See also §13002 of the Dickey Act, which referred to "the use of water for any beneficial use other than the use for disposal of sewage and industrial wastes."

17. See §5412 of the final Act (A.B. 2156). A June 22, 1949, analysis sent by the Department of Agriculture to James H. Oakley, Governor Warren's executive secretary, says that "what Assembly Bill No. 2156 is designed to accomplish…appears to be quite well known; namely, to place a limitation on the powers of the State Department of Public Health with reference to the pollution problem,…"

18. §13154 of A.B. 2034; A.B. 2034, as proposed before June 15, §§13053, 13064; memorandum from Director of Public Works to Governor Warren, June 21, 1949 (Dic), pp. 6–7. In the original A.B. 2156 introduced by Dickey on January 27, 1949, jurisdiction over waste disposal remained in the Health Department. The bill was amended on April 28 to transfer jurisdiction to newly created Regional Boards.

19. A.B. 2156 (as originally introduced), §§5410(e), 5414, 5410(f), 5415, 5410(g), 5416. The chemical industry's opposition to general standards is discussed by Smith, "Silk Purses into Sows' Ears."

20. "Groups Denounce Dickey Water Bill," *Los Angeles Times,* June 16, 1949; Hanson, "Water Pollution Action Postponed." Full text of the statement of opposition was printed in the *Assembly Journal,* June 16, 1949, pp. 4585–4586. League of Cities and sewage treatment funding: E. Cray, *Chief Justice* (Simon & Schuster, New York, 1997), pp. 169–170.

21. The clearest description of the effect of the amendments is in the Director of Public Works June 21 memorandum, pp. 11–13. "Pollution Issue Nears Compromise," *Los Angeles Times,* June 17, 1949, is weak on the details of the compromise.

22. A.B. 2034, §13062–3. The committee that revised the Dickey Act in 1969 reported that enforcement procedures had been criticized as weak and ineffective because "trial courts.... have refused to grant injunctive relief where the trial judge was not satisfied with the reasonableness of regional board discharge requirements." H. O. Banks et al., "Preliminary Report of the Study Panel to the California State Water Resources Control Board," January 1969, p. 9.

23. Memorandum, L. C. Larson to Regional Water Pollution Board No. 4, April 20, 1953 (Aut); Santa Ana Regional Water Pollution Control Board Resolutions No. 53–5 and 53–6, 1953 (San).

24. J. E. McKee, *Water Quality Criteria,* State Water Pollution Control Board Publication No. 3, 1952, pp. 4–6; E. J. Cleary, "Jack Edward McKee," in *Memorial Tributes: National Academy of Engineering, Volume 2* (National Academies Press, Washington, 1984), pp. 204–209.

25. McKee, *Water Quality Criteria;* J. E. McKee, "The Need for Water Quality Criteria," *Proc. Physiological Aspects of Water Quality,* Washington, September 8–9, 1960.

26. McKee, *Water Quality Criteria,* pp. 5, 12, 26, 214–217. McKee does make passing mention of the carcinogenicity of arsenic and radioactivity.

27. McKee, "Need for Water Quality Criteria"; McKee, *Water Quality Criteria,* p. 12.

28. §13264(a).

29. H. O. Banks et al., *Recommended Changes in Water Quality Control,* State of California, The Resources Agency, March 1969, Appendix A, p. 26, quoted the existing state code and commented that "After careful consideration, the portion of the quoted definition calling waste disposal, etc., 'economic beneficial uses of water,' was not included in the proposed new definition."

30. §§13300, 13304.

31. Perlmutter, Lieber, and Frauenthal, "Movement of Waterborne Cadmium and Hexavalent Chromium Wastes"; H. W. Davids and M. Lieber, "Underground Water Contamination by Chromium Wastes," *Water and Sewage Works,* vol. 98, pp. 528–534 (1951).

32. *Progress Report of the Special Committee on Pollution Abatement of the Joint Legislative Committee on Interstate Cooperation,* Legislative Document No. 59 (Williams Press, Albany, 1947), especially pp. 27–29; *Report of the Special Committee on*

Pollution Abatement of the Joint Legislative Committee on Interstate Cooperation, Legislative Document No. 51 (Williams Press, Albany,1949), esp. pp. 23, 25–27; *Report of the Special Committee on Pollution Abatement of the Joint Legislative Committee on Interstate Cooperation,* Legislative Document No. 53 (Williams Press, Albany,1950), esp. p. 19.

33. *Progress Report of the Special Committee on Pollution Abatement,* pp. 32–40. The similarities between the provisions criticized by Ostertag and those that hobbled enforcement of California's water pollution controls under the Dickey Act are so striking that one wonders whether Dickey, aided by industry's systematic exchange of information among states, drew a perverse inspiration from the 1947 New York report.

34. *Progress Report of the Special Committee on Pollution Abatement,* p. 9; "State Board Urged on Water Pollution," *New York Times,* Sept. 5, 1947; "Fast Action Seen on Pollution Bill," *New York Times,* Jan. 22, 1949; "Albany Gets Bill on Water Spoilage," *New York Times,* March 1, 1949; "Dewey Asks Action to Cut Pollution," *New York Times,* March 4, 1949; "Curb on Pollution Is Sent to Dewey," *New York Times,* March 25, 1949. The legislative history contains petitions in favor sent to the governor by a variety of fishing, garden, and conservation clubs, along with a Feb. 3 note from M. F. Hilfinger, president of the Associated Industries of New York State, to Charles D. Breitel, the Governor's counsel, asking for the chance to discuss "at least one particular feature" of the proposed bill.

35. *Report of the Special Committee on Pollution Abatement* (1950), pp. 27–30, 70–87; H. E. Babbitt, "The Administration of Stream Pollution Prevention in Some States," in *Proceedings of the Sixth Industrial Waste Conference* (Purdue University, Lafayette, IN, 1951), pp. 239–252; McKee, *Water Quality Criteria,* pp. 423–431.

36. Travis, *Dyes Made in America,* pp. 377–386.

37. Andreen, "Water Pollution Control: Part I," esp. pp. 193–197; Brook, *Before Earth Day,* pp. 54–57; Kehoe, *Cleaning Up the Great Lakes,* p. 30. McKee, *Water Quality Criteria,* pp. 59–96, summarizes the regulations of numerous states and interstate commissions.

38. C. E. Colten, "Groundwater and the Law: Records v. Recollections," *Public Historian,* vol. 20, pp. 25–44 (1998); Kehoe, *Cleaning Up the Great Lakes,* pp. 25–26; Babbitt, "Administration of Stream Pollution Prevention"; *Control of Ground Water,* Illinois Legislative Council Publication 88 (Springfield, 1948); 1951 Illinois Revised Statutes, Chap. 10, §129–145; Alabama Act 523, 1947 Regular Session; Act 463, 1953 Regular Session.

39. Kehoe, *Cleaning Up the Great Lakes,* pp. 29–36.

Chapter 11

1. R. W. Brenneman, "Evaporation Pond, Plant B-1 Industrial Waste Disposal Facility," undated Lockheed Aircraft Corp. document (Aut).

2. R. E. Doherty, "A History of the Production and Use of Carbon Tetrachloride, Tetrachloroethylene, Trichloroethylene and 1,1,1-Trichloroethane in the United States: Part 1—Historical Background; Carbon Tetrachloride and Tetrachloroethylene," *Environmental Forensics,* vol. 1, pp. 69–81 (2000); R. E. Doherty, "A History of the Production and Use of Carbon Tetrachloride, Tetrachloroethylene, Trichloroethylene and 1,1,1-Trichloroethane in the United States: Part 2—Trichloroethylene and 1,1,1-Trichloroethane," *Environmental Forensics,* vol. 1, pp. 83–93 (2000); R. D. Morrison and B. L. Murphy, "Chlorinated Solvents: Chemistry, History and Utilization for Source Identification and Age Dating," in B. L. Murphy and R. D. Morrison, eds., *Introduction to Environmental Forensics* (Academic Press, San Diego, 2002), pp. 261–310. Chloromethane: See chapter 4 above.

3. Doherty, Part 2; Brienes, "Fight against Smog," pp. 183–184. Brienes refers to "chlorinated hydrocarbons," a term that encompasses both the chlorinated solvents and chlorinated pesticides such as DDT, but the chlorinated solvents are much more volatile and are undoubtedly what the APCD had in mind.

4. J. E. Edwards, "Hepatomas in Mice Induced with Carbon Tetrachloride," *Journal of the National Cancer Institute,* vol. 2, pp. 197–199 (1941); A. B. Eschenbrenner, "Induction of Hepatomas in Mice by Repeated Oral Administration of Chloroform, with Observations on Sex Differences," *Journal of the National Cancer Institute,* vol. 5, pp. 251–255. Short-handed: Russell, "Strange Career of DDT."

5. Hueper, "Environmental Cancer," p. 6; "Environmental and Occupational Cancer," pp. 21–22.

6. Doherty, "A History: Part I."

7. J. A. Zapp, Jr., "Evaluation Procedures in Use at the Haskell Laboratory of Industrial Toxicology," Nov. 19, 1951 (Aut).

8. F. A. Lyne and T. McLachlan, "Contamination of Water by Trichloroethylene," *The Analyst,* vol. 74, p. 513 (1949).

9. S. Amter and B. Ross, "Discussion of 'A Quest to Locate Sites Described in the World's First Publication on Trichlorethene Contamination of Groundwater' by M. O. Rivett and L. Clark," *Quarterly Journal of Engineering Geology and Hydrology,* vol. 41, pp. 491–493 (2008). A contrary view of the reaction to this paper is found in M. O. Rivett, S. Feenstra, and L. Clark, "Lyne and McLachlan (1949): Influence of the First Publication on Groundwater Contamination by Trichloroethene," *Environmental Forensics,* vol. 7, pp. 313–323 (2006) and M. O. Rivett and L. Clark, "A Quest to Locate Sites Described in the World's First Publication on Trichlorethene Contamination of Groundwater," *Quarterly Journal of Engineering Geology and Hydrogeology,* vol. 40, pp. 241–249 (2007).

10. H. O. Banks and J. H. Lawrence, "Water Quality Problems in California," *Transactions of the American Geophysical Union,* vol. 34, no. 1, pp. 58–66, 1953. Although chlorinated solvents are not specifically mentioned in the paper, they are members of the broader category of "organic solvents." See Amter and Ross, "Discussion of Rivett and Clark."

11. RWPCB, Central Valley Region, *Resolution No. 127*, May 15, 1952.

12. Inter-Departmental Communication from P. J. Coffey, DWR Supervising Hydraulic Engineer, to J. S. Gorlinski, Executive Officer Central Valley RWPCB, dated January 10, 1955, subject: Waste discharge, Aerojet-General Plant, Nimbus, Sacramento County (Aut). Besides TCE and PCE, the letter mentions the unchlorinated solvents butyl and amyl acetate, methyl isobutyl ketone, and xylene.

13. A. L. Taylor, "Nematocides and Nematicides—A History," *Nematropica,* vol. 33, pp. 225–232 (2003).

14. See for example R. P. Daroga and A. G. Pollard, "Colorimetric Method for the Determination of Minute Quantities of Carbon Tetrachloride and Chloroform in Air and in Soil," *Journal of the Society of the Chemical Industry,* vol. 60, pp. 218–222 (1941), whose tests of the use of carbon tetrachloride and chloroform as fumigants represents one of the earliest investigations of the migration of chlorinated solvent compounds in geologic soil materials. Also, C. T. Schmidt, "Dispersion of Fumigants through Soil," *Journal of Economic Entomology,* vol. 40, pp. 829–837 (1947), investigated the dispersion of fumigants through soil. W. J. Hanson, "Factors Which Influence the Diffusion of Fumigants through Soil," Dow Chemical Company, Oct. 6, 1950 (Aut); C. R. Youngson, "The Effects of Some Soil Environmental Conditions on Movement of and Root-knot Nematode Control by 1,2-dibromo-3-chloro-propane, 3-bromopropene and 1,3-dichloropropene," Dow Chemical Company, Oct. 26, 1956; Dow Agricultural Chemical Development, "Factors Influencing Diffusion and Nematode Control by Soil Fumigants," ACD Information Bulletin No. 110, Nov. 29, 1957 (Aut). D. H. Smith and R. S. Shigenaga, "Extraction of Fumigants from Soil for Their Determination by Gas-Liquid Chromatography," *Soil Science Society of America Proceedings,* vol. 25, pp. 160–161 (1961).

15. *Western Fruit Grower,* April 1961; J. Collins and A. W. Feldman, "Penetration of Nematocides for Control of *Radopholus similis* and for Destruction of Citrus Roots in the Deep Sands of Central Florida," *Phytopathology,* vol. 55, pp. 1103–1107 (1965).

16. Shell Technical Service Laboratory Technical Memoranda 55/65 and 63/65, "Residues of DD in Samples of Water from France," March 24, 1965 and April 9, 1965 (Aut); E. F. Feichtmeir, Shell, letter to R. F. Woodward, Shell, Nov. 11, 1965 [the date appears to be an error] (Aut).

17. C. A. I. Goring, "Fumigants, Fungicides, and Nematicides," in C. A. I. Goring and J. W. Hamaker, *Organic Chemicals in the Soil Environment* (Marcel Dekker, New York, 1972), p. 591.

18. Dickey Report, attached table. Manufacturing of 2,4-D is the only branch of synthetic organic chemical production included in the table; the industry chose not to call attention to the problems of other products' manufacturing processes. Montebello: Swenson, "The Montebello Incident."

19. Manufacturing Chemists Association, minutes of Legal Advisory Committee, Sept. 15, 1949 (MCA).

20. Manufacturing Chemists Association of the United States, *Trichloroethylene, Chemical Safety Data Sheet SD-14,* 1947.

21. J. F. Pankow, S. Feenstra, J. A. Cherry, and M. C. Ryan, "Dense Chlorinated Solvents in Groundwater: Background and History of the Problem," in *Dense Chlorinated Solvents and Other DNAPLs in Groundwater,* J. F. Pankow and J. A. Cherry, eds. (Waterloo Press, Portland, 1995), pp. 1–52; R. E. Jackson, "Anticipating Ground-Water Contamination by New Technologies and Chemicals: The Case of Chlorinated Solvents in California," *Environmental and Engineering Geoscience,* vol. 5, pp. 331–338 (1999); R. E. Jackson, "Chlorinated Solvents and the Historical Record: A Response to Amter and Ross," *Environmental Forensics,* vol. 4, pp. 3–9 (2003). We have responded to these arguments in detail in S. Amter and B. Ross, "Was Contamination of Southern California Groundwater by Chlorinated Solvents Foreseen?" *Environmental Forensics,* vol. 2, pp. 179–184 (2001).

22. Amter and Ross, "Was Contamination Foreseen?"

Chapter 12

1. L. Cox, "Company Investment in Pollution Abatement Facilities" (memorandum to M. F. Wood, Aug. 14, 1950) (Hag). Cox was a leading water pollution expert at DuPont and represented the company on the Manufacturing Chemists' Association water pollution committee.

2. MCA presidents 1873–1943, typed list (CHF); MCA Executive Committee minutes, June 4, 1942; May 14, 1946 (MCA).

3. Fortunately, the company's record is illuminated by a wide variety of internal documents. DuPont established a nonprofit archive, the Hagley Museum and Library, where many of its archives were made available to scholars, and numerous researchers have taken advantage of this resource.

4. Smith, "Silk Purses into Sows' Ears."

5. Smith, "Silk Purses into Sows' Ears"; Chandler and Salisbury, *Pierre S. du Pont,* chaps. 6 and 8.

6. Hounshell and Smith, *Science and Corporate Strategy,* pp. 337–346.

7. M. S. Gerber, *A Brief History of the T Plant Facility at the Hanford Site,* Westinghouse Hanford Report WHC-MR-0452, Addendum 1, May 1994, pp. 23–33; M. S. Gerber, *On the Home Front* (University of Nebraska Press, Lincoln, 1992), pp. 34–37, 77–86.

8. Hounshell and Smith, *Science and Corporate Strategy,* pp. 338–345, 445–447, 450–451; Russell, *War and Nature,* p. 147.

9. Gerber, *T Plant,* pp. 23–24.

10. Ndiaye, *Nylon and Bombs,* pp. 197, 208–209; National Research Council, *Tank Waste Retrieval, Processing, and On-site Disposal at Three Department of Energy Sites* (National Academy Press, Washington, 2006), pp. 13–19.

11. Ndiaye, *Nylon and Bombs,* pp. 172–178, 186.

12. G. M. Read, "Control and Abatement of Pollution" (M. F. Wood's copy, with handwritten notes, of memorandum to Executive Committee, March 9, 1949); F. G. Kess, "Control and Abatement of Pollution" (advice of action, March 23, 1949) (Hag). For 1930s, see also Smith, "Silk Purses into Sows' Ears."

13. L. Cox, memo to M. F. Wood, Aug. 14, 1950 (Hag); Hounshell and Smith, *Science and Corporate Strategy,* pp. 98, 328.

14. A classic analysis of the role that control of information plays when a complex economic unit is centrally managed is J. Kornai, *The Socialist System* (Princeton University Press, Princeton, 1992).

15. G. M. Read, "Summary of Figure Data—Industrial Department Reports on Water and Air Pollution" (memo to Executive Committee, Aug. 9, 1950) (Hag). The May 16 summary report has not come to light, but it is cited by G. F. Jenkins, Bibliography, in *Air Pollution Abatement Manual* (Manufacturing Chemists Association, 1952), p. 56.

16. Hounshell and Smith, *Science and Corporate Strategy,* p. 211.

17. Read, Summary of figure data, Aug. 9, 1950; Grasselli Board of Directors minutes, Oct. 22, 1929 (Hag).

18. Hounshell and Smith, *Science and Corporate Strategy,* pp. 212–214, 217–218.

19. E. C. Thompson, "Sale of Meadowbrook and New Castle Plants" (memorandum to Executive Committee, Aug. 9, 1950) (Aut). This memorandum says that the Grasselli Department had been trying to sell since it was instructed to do so by a 1943 resolution of the Executive Committee and had failed to sell only for lack of a willing buyer, but Grasselli had ignored the further instruction in the 1943 resolution to close the plant as soon as the war ended if no buyer was found.

20. Read, "Control and Abatement of Pollution"; Summary of figure data, Aug. 9, 1950. Location of the El Monte plant, between the current El Monte Bus Terminal and Interstate 10, is from an interview with Donna Crippen of the El Monte Historical Museum, July 29, 2008, and confirmed by a Du Pont Access Drive shown on the Environmental Impact Statement for the bus terminal. The sewer outfall in the Montebello incident is located on the Rio Hondo one mile upstream of Whittier Narrows by Swenson, "The Montebello Incident"; the former DuPont location adjoins the Rio Hondo three miles upstream of Whittier Narrows.

21. L. Cox, "Company Investment in Pollution Abatement Facilities" (memorandum to M. F. Wood, August 14, 1950) (Hag).

22. H. W. de Ropp, "Chemical Waste Disposal at Victoria, Texas, Plant of the du Pont Company, *Sewage and Industrial Wastes,* vol. 23, pp. 194–197 (1951); H. O. Henkel, "Surface and Underground Disposal of Chemical Wastes at Victoria, Texas," *Sewage and Industrial Wastes,* vol. 25. pp. 1044–1049 (1953). Earlier chemical company efforts: L. K. Cecil, "Underground Disposal of Process Waste Water," *Industrial and Engineering Chemistry,* vol. 42, pp. 594–599 (1950).

23. R. S. Karpiuk, *Dow Research Pioneers: Recollections* (Pendell Publishing, Midland MI, 1984), pp. 612–613.

24. S. Lenher, Opening remarks, air and water resources meeting, June 12, 1969 (Hag).

Chapter 13

1. Minutes, Air Pollution Abatement Committee, Manufacturing Chemists' Association, Feb. 18, 1954 (MCA).

2. Colten and Skinner, *Road to Love Canal,* pp. 105–106; Smith, "Silk Purses into Sows' Ears."

3. Colten and Skinner, *Road to Love Canal,* pp. 55–58, 105–106; R. W. Hess, "Wastes from Chemical Manufacturing," *Sewage Works Journal,* vol. 21, pp. 674–682 (1949); discussion by T. J. Powers, p. 683; Minutes of joint meeting, Stream Pollution Abatement Committee and Legal Advisory Committee, July 13, 1949 (MCA).

4. Minutes, Executive Committee, Manufacturing Chemists' Association, Sept. 9, 1947 and Oct. 14, 1947; Board of Directors, Jan. 11, 1949, April 12, 1949, and Sept. 13, 1949; Air Pollution Abatement Committee, Nov. 9, 1949, Jan. 11, 1950, Sept. 13, 1950, Dec. 12, 1951 (MCA).

5. Interviews with nieces Donna Gosline, July 11, 2008, and Anne Barnes (quotation), July 14, 2008; Anne Barnes, written communication, Nov. 15, 2008; "Industry Advised on Air Pollution," *New York Times,* Feb. 27, 1952.

6. "George E. Best, Retired Chemist, Dies," *Baltimore Sun,* Oct. 8, 1990; "Grant Is Elected T. C. A. President, Rucker Treasurer," *The Tech,* March 2, 1934; Hartford, "Fells Point"; O. F. Tarr, memo to J. Benington, Feb. 26, 1948 (ICO); interview with Joyce Zimmerman (church acquaintance of Best), Aug. 18, 2009. Home of Best's wife: http://www4.esu.edu/admissions/scholarships_fa/loans_grants.cfm, accessed July 29, 2008.

7. Minutes, Air Pollution Abatement Committee, Nov. 9, 1949 (MCA).

8. Minutes, Air Pollution Abatement Committee, Dec. 7, 1950; April 17, 1951; Oct. 18, 1951; Oct. 22, 1952; April 22, 1953; Sept. 18, 1953; Sept. 29, 1954 (MCA). *Air Pollution Abatement Manual,* Manufacturing Chemists Association, Manual Sheets P-1 through P-12, 1951–1954.

9. Minutes, Air Pollution Abatement Committee, Sept. 13, 1950; Dec. 7, 1950; March 14, 1951; April 17, 1951; May 17, 1951; Oct. 18, 1951; Dec. 12, 1951. G. E. Best, "A Rational Approach to Air Pollution Legislation," *Industrial Hygiene Quarterly,* vol. 13, pp. 62–69 (1952).

10. Quotation from minority report attached to minutes of Air Pollution Abatement Committee, Manufacturing Chemists Association, May 17, 1951. That the minority report represented the position of industry is stated in the minutes of Oct. 18, 1951 (MCA).

11. Minutes, Air Pollution Abatement Committee, Manufacturing Chemists Association, March 14, 1951 (MCA).

12. Minutes, Air Pollution Abatement Committee, Manufacturing Chemists Association, Jan. 11, 1950 (MCA).

13. Minutes, Air Pollution Abatement Committee, Manufacturing Chemists Association, Jan. 11, 1950; Minutes of joint meeting on 1951 pollution abatement conference, Jan. 10, 1951 (MCA); "Polluted Air Held No Health Menace," *New York Times,* Feb. 26, 1952. Establishment of public relations committee: MCA Executive Committee minutes, April 9, 1946. See also "Industry Advised in Air Pollution," *New York Times,* Feb. 27, 1952: "Plant diseases and other non-industrial factors sometimes cause crop damage for which blame is wrongly put on air pollution by industry, the Manufacturing Chemists Association was told yesterday."

14. Minutes, Air Pollution Abatement Committee, Feb. 26–27, 1953; April 22, 1953; Feb. 18, 1954 (MCA).

15. Minutes, Air Pollution Abatement Committee, Feb. 18, 1954 (MCA); N. F. Billings et al., "Control of Underground Waste Disposal," *Journal of the American Water Works Association,* vol. 44, pp. 685–689 (1952). Powers: E. N. Brandt, *Growth Company: Dow Chemical's First Century* (Michigan State University Press, East Lansing, 1997), pp. 265–266; Karpiuk, *Dow Research Pioneers,* pp. 607–612. The minutes of MCA's water pollution committee from this period have not, with a few exceptions, come to light; the comment here by Powers suggests that they would be worth examining.

16. Hartford, "Fells Point"; Minutes, Board of Directors, Manufacturing Chemists Association, Dec. 8, 1959; Minutes, Executive Committee, MCA, October 10, 1978 (MCA); "George E. Best, Retired Chemist, Dies."

17. Spitz, *Petrochemicals,* pp. 197–226, 302, 307–310, 338–342; Travis, *Dyes Made in America,* pp. 91–94, 137–138, 148.

18. Colten and Skinner, *Road to Love Canal,* pp. 48–63, 93–94; Smith, "Silk Purses into Sows' Ears."

Chapter 14

1. J. C. Whitaker, "Earth Day Recollections: What It Was Like When the Movement Took Off," *EPA Journal,* vol. 14, no. 8 (1988), online at http://www.epa.gov/history/topics/earthday/10.htm.

2. Stradling, *Smokestacks and Progressives,* pp. 186–188.

3. Snyder, "Death-Dealing Smog," pp. 237–240.

4. Davis, *When Smoke Ran Like Water,* pp. 42–54.

5. Dewey, *Don't Breathe the Air,* p. 246; Brooks, *Before Earth Day,* pp. 134–145; G. Hill, "U.S. Action Urged in Smog Problem," *New York Times,* April 21, 1955; B. Furman, "$25,000,000 Study of Smog Proposed," *New York Times,* June 12, 1955; Minutes of Air Pollution Abatement Committee, Manufacturing Chemists' Association, April 21, 1954, Sept. 29, 1954, and Feb. 9, 1955 (MCA); "McCabe to Take Office as U.S. Smog Adviser," *Los Angeles Times,* April 30, 1955.

6. Herzog, "Louis C. McCabe"; "McCabe to Take Office"; "Louis McCabe Quits as PHS Expert," *Washington Post,* Jan. 5, 1956. Anti-regulatory bias: Dewey, *Don't Breathe The Air,* pp. 226–227, 244.

7. Bosso, *Pesticides and Politics,* pp. 71–78; Simon, *DDT,* pp. 138–141.

8. Bosso, *Pesticides and Politics,* pp. 67–71; Simon, *DDT,* pp. 137, 142–144; P. Daniel, *Toxic Drift: Pesticides and Health in the Post-World War II South* (Louisiana State University Press, Baton Rouge, 2005), pp. 5–13, 55–66, 101–106, 130–132, 157–158; Taylor, "Nematocides and Nematicides."

9. Bosso, *Pesticides and Politics,* pp. 79–106; Simon, *DDT,* pp. 144–145; Daniel, *Toxic Drift,* pp. 55–61.

10. R. Carson, *Silent Spring* (Houghton Mifflin, Boston, 1962); Bosso, *Pesticides and Politics,* pp. 115–120; Proctor, *Cancer Wars,* pp. 48–53; Simon, *DDT,* pp. 146–163; Daniel, *Toxic Drift,* pp. 66–70; "Rachel Carson's Warning," *New York Times,* July 2, 1962.

11. Bosso, *Pesticides and Politics,* pp. 120–132; Daniel, *Toxic Drift,* pp. 70–79.

12. Warren, *Brush with Death,* pp. 210–219; Nriagu, "Clair Patterson"; Markowitz and Rosner, *Deceit and Denial,* pp. 108–117.

13. McCulloch and Tweedale, *Defending the Indefensible,* pp. 84–96.

14. Proctor, *Cancer Wars,* pp. 45–46; J. D. Dingell, letter to L. L. Terry, Aug. 29, 1961 (NCI); Breslow, *History of Cancer Control.* Dingell was a persistent critic of the PHS; he also took the first initiative in 1959 to end its authority over water pollution control. Milazzo, *Unlikely Environmentalists,* p. 59.

15. Andreen, "Water Pollution Control: Part II"; Milazzo, *Unlikely Environmentalists,* pp. 22, 29–30, 33–34, 58.

16. Dewey, *Don't Breathe the Air,* pp. 238–239.

17. Milazzo, *Unlikely Environmentalists,* pp. 61–85; Andreen, "Water Pollution Control: Part II"; Kehoe, *Cleaning Up the Great Lakes,* pp. 76–83.

18. Jones, *Clean Air,* pp. 58–84, 108–123; Dewey, *Don't Breathe the Air,* pp. 240–244.

19. K. C. Clarke and J. J. Hemphill, "The Santa Barbara Oil Spill, a Retrospective," *Yearbook of the Association of Pacific Coast Geographers,* vol. 64, pp. 157–162 (2002); J. H. Adler, "Fables of the Cuyahoga: Reconstructing a History of Environmental Protection," *Fordham Environmental Law Journal,* vol. 14, pp. 89–146 (2003); Kehoe, *Cleaning Up the Great Lakes,* pp. 134–135.

20. P. Shabecoff, *A Fierce Green Fire: The American Environmental Movement* (Hill and Wang, New York, 1993), pp. 113–119; Milazzo, *Unlikely Environmentalists,* pp. 146–150; Jones, *Clean Air,* pp. 145–146; Dewey, *Don't Breathe the Air,* p. 250.

21. D. P. Moynihan, *The Politics of a Guaranteed Income: The Nixon Administration and the Family Assistance Plan* (Vintage, New York, 1973), p. 215. Milazzo, *Unlikely Environmentalists,* pp. 150–152, offers a shrewd assessment of Nixon's approach to the environment.

22. Jones, *Clean Air*, pp. 179–210; Dewey, *Don't Breathe the Air*, p. 243; Milazzo, *Unlikely Environmentalists*, pp. 133, 140, 153–157; J. Quarles, *Cleaning Up America* (Houghton Mifflin, Boston, 1976), p. 56; Hollander, *Abel Wolman*, p. 164.

23. Milazzo, *Unlikely Environmentalists*, pp. 166–173.

24. Milazzo, *Unlikely Environmentalists*, pp. 174–176, 191–237, 247–248; Andreen, "Water Pollution Control: Part II."

25. Andreen, "Water Pollution Control: Part II."

26. Bosso, *Pesticides and Politics*, pp. 154–177.

27. T. P. Wagner, "Hazardous Waste: Evolution of a National Environmental Problem," *Journal of Policy History*, vol. 16, pp. 306–331 (2004); "Report to Congress: Disposal of Hazardous Wastes," USEPA Report SW-115, 1974, pp. 1, 7. EPA Administrator Ruckelshaus told a Senate committee in March 1971 that "For the overwhelming majority of elements or compounds that might be included as hazardous substances..., there is generally insufficient verified information...to permit the establishment of definitive effluent limitations other than total prohibition." *Water Pollution Control Legislation*, Hearings of the Subcommittee on Air and Water Pollution, Committee on Public Works, U.S. Senate, March 15–24, 1971, p. 81.

28. Wagner, "Hazardous Waste." PCBs: C. Q. Gustafson, "PCBs—Prevalent and Persistent," *Environmental Science and Technology*. vol. 4, pp. S14–S19 (1970); "U.S. Aides Seek to Calm Fears over Chemical PCB," *New York Times*, Sept. 30, 1971; *Polychlorinated Biphenyls and the Environment*, Interdepartmental Task Force on PCBs Report COM-72-10419, May 1972.

29. Wagner, "Hazardous Waste."

30. M. Middleton and G. Walton, "Organic Contamination of Ground Water," pp. 50–56; L. M. Miller, "Contamination by Processed Petroleum Products," pp. 117–119, and "Discussion 1," pp. 32 and 33, in *Ground Water Contamination: Proceedings of the 1961 Symposium*, Technical Report W61–5 (Robert A. Taft Sanitary Engineering Center, Cincinnati, 1961); D. W. Miller, F. A. Deluca, and T. L. Tessier, *Ground Water Contamination in the Northeastern States*, USEPA Report EPA-660/2–74–056, June 1974, pp. 155, 233, 234. Numerous citations of Lyne and McLachlan are listed in S. Amter and B. Ross, "Discussion of 'A Quest to Locate Sites Described in the World's First Publication on Trichloroethene Contamination of Groundwater' by M. O. Rivett & L. Clark," *Quarterly Journal of Engineering Geology and Hydrogeology*, vol. 41, pp. 491–497 (2008).

31. *Water Pollution Control Legislation*, Hearings of the Subcommittee on Air and Water Pollution, Committee on Public Works, U.S. Senate, April 5, 1971, p. 3205.

32. Rosner and Markowitz, *Deceit and Denial*, pp. 181–224; Davis, *Secret History*, pp. 368–379; G. E. Best, letter to J. D. Bryan, March 26, 1973, http://www

.deceitanddenial.org/gallery/ScrantonDocuments/MoralOb11973, accessed Aug. 4, 2008.

33. Markowitz and Rosner, *Deceit and Denial,* pp. 204–220; H. M. Schmeck, Jr., "Water from Mississippi River Linked to Cancer Death Trends," *New York Times,* Nov. 8, 1974.

34. J. F. Pankow, S. Feenstra, J. A. Cherry, and M. C. Ryan, "Dense Chlorinated Solvents in Groundwater: Background and History of the Problem," in *Dense Chlorinated Solvents and Other DNAPLs in Groundwater,* J. F. Pankow and J. A. Cherry, eds. (Waterloo Press, Portland, 1995), pp. 1–52; J. M. Symons et al., "National Organics Reconnaissance Survey for Halogenated Organics," *Journal of the American Water Works Association,* vol. 67, pp. 634–647 (1975).

35. *Ground Water Quality Research & Development,* Hearings of the Committee on Science and Technology, U.S. House of Representatives, April 8, 26, 27, 1978, pp. 31–32, 36–37, 43, 79.

36. *Ground Water Quality,* pp. 23, 36–44, 79.

37. T. A. Burke, oral communication, July 22, 2009; T. A. Burke and R. K. Tucker, *A Preliminary Report on the Findings of the State Groundwater Monitoring Project,* New Jersey, Dept. of Environmental Protection, March 1978; *Ground Water Quality,* p. 32.

38. *Ground Water Quality,* p. 151.

39. S. Weart, *The Discovery of Global Warming* (Harvard University Press, Cambridge, 2008), available online at www.aip.org/history/climate; T. C. Peterson, W. M. Connelley, and J. Fleck, "The Myth of the 1970s Global Cooling Scientific Consensus," *Bulletin of the American Meteorological Society,* vol. 89, pp. 1325–1337 (2008).

40. "Elon Hooker Dies, Manufacturer, 68," *New York Times,* May 11, 1938.

41. A. Mazur, *A Hazardous Inquiry: The Rashomon Effect at Love Canal* (Harvard University Press, Cambridge, 1999), pp. 8–10; Colten and Skinner, *Road to Love Canal,* pp. 151–154.

42. Mazur, *A Hazardous Inquiry,* pp. 11–15; J. A. Tarr and C. Jacobson, "Environmental Risk in Historical Perspective," in B. B. Johnson and V. T. Covello, *The Social and Cultural Construction of Risk* (Reidel, Dordrecht, 1988), pp. 317–344; Wagner, "Hazardous Waste."

43. Wagner, "Hazardous Waste"; W. H. Frank and T. B. Atkeson, *Superfund: Litigation and Cleanup* (Bureau of National Affairs, Washington, 1985). The formal name of the law is the Comprehensive Environmental Response, Compensation, and Liability Act, or CERCLA.

Chapter 15

1. C. K. Banks, "The Interdependence of Man and Earth," *Proc. Southern Industrial Wastes Conference,* Houston, April 21–23, 1954, pp. 8–29.

2. A. P. Black, "A Rational Approach to the South's Pollution Problem," *Proc. Southern Industrial Wastes Conference,* Houston, April 21–23, 1954, pp. 1–7;

Banks, "Interdependence of Man and Earth." Black: "Dr. A. P. Black: An Environmental Engineering Scientist," http://www.ees.ufl.edu/apblack/intro.asp, accessed Sept. 8, 2008. Banks: C. K. Banks, Jr., written communication, Sept. 10, 2008.

3. Black, "Rational Approach."

4. Banks, "Interdependence of Man and Earth." Burt: Manufacturing Chemists' Association, minutes, Board of Directors, May 11, 1954.

INDEX

Adams, Ed, 148–149
adiponitrile, 138
Agriculture Department, 46–48, 153–155
  Bureau of Entomology, 55–56, 121–122
  Insecticide Division, 55, 57–58
  See also Food and Drug Administration
Ainsworth, Ed, 81
Air Hygiene Foundation. *See* Industrial
  Hygiene Foundation
air pollution
  from chemical plants, 66, 70, 119,
    133–137, 143–149
  coal smoke, 9–10, 73–74
  in Donora, Pennsylvania, 1–2, 86–97
  legislation, 95–96, 146–148, 151–153,
    156–159, 169
  in Los Angeles, 73–85, 96–97, 119
  radioactive, 131
  from smelting, 10–13, 136–137
  See also leaded gasoline
Air Pollution and Smoke Prevention
  Association of America, 147,
    151–152
Air Pollution Control Act (1955), 5, 152, 162
Alabama Water Improvement Commission,
  116
Alcoa Aluminum, 66, 143, 147–148
Alexander, Hope, 88
Alexandria (Virginia), 30, 41–42
Allied Chemical and Dye, 25, 72, 149,
  150

American Chemical Society, 6, 19,
  105–106, 143
American Cyanamid, 18, 21, 23–25, 142,
  150
American Defense Society, 23
American Institute of Mining and
  Metallurgical Engineers, 37
American Liberty League, 22, 24, 27, 103
American Medical Association, 60
American Petroleum Institute, 14–16, 99,
  102–103, 142–143
  Manual of Disposal of Refinery Wastes, 15, 105
American Smelting and Refining
  Corporation, 12
American Water Works Association, 149
ammonia, 116
Anaconda Copper Company, 11–12
Analyst, The, 120
Arkansas, 13
Army, U.S., 51–52, 54–55, 104, 161
  Corps of Engineers, 14–15, 99, 159
  Surgeon General, 54, 119, 132
Arrhenius, Svante, 164, 45–47, 49–51, 53,
  86
arsenic
  as air pollutant, 10–13, 86, 147
  as pesticide, 12, 45–51, 53, 57
asbestos, 35, 64, 66, 92, 156
aspirin, 18
Atlantic Refining Company, 105
Atomic Energy Commission, 67